James Adams is the defence correspondent of *The Sunday Times* and a regular contributor to the *Washington Post* and the *Los Angeles Times*. He is the author of *The Unnatural Alliance*, which dealt with relations between Israel and South Africa, *The Financing of Terror* and *Secret Armies*, the story of covert warfare. He has lectured extensively in Europe and America on terrorism and covert warfare.

Robin Morgan is the features editor of *The Sunday Times* and a former editor of the Insight investigative team. He won Campaigning Journalist of the Year in the 1983 British Press Awards and has co-authored *The Falklands War: The Full Story, Rainbow Warrior: The French Attempt to Sink Greenpeace* and *Bullion*, the story of Britain's biggest gold robbery.

Anthony Bambridge is associate editor of *The Sunday Times* and is also responsible for the newspaper's long-term investigations. He has worked extensively on quality British newspapers, holding a wide range of executive positions and is a former news editor of *The Observer* and writer for the *Economist*.

James Adams
Robin Morgan
Anthony Bambridge

AMBUSH

The War Between the SAS and the IRA

Pan Original
Pan Books London, Sydney and Auckland

First published 1988 by Pan Books Ltd,
Cavaye Place, London SW10 9PG

9 8 7 6 5 4 3 2 1

© James Adams, Robin Morgan and Anthony Bambridge 1988

ISBN 0 330 30893 9

Photoset by Parker Typesetting Service, Leicester
Printed and bound in Great Britain by
Richard Clay Ltd, Bungay, Suffolk

Acknowledgements

This book would not have been possible without the unstinting co-
operation of our colleagues, contacts and friends. There are many we
cannot name, for their positions demand anonymity. We can, however,
thank Andrew Neil, the editor of *The Sunday Times*, for his support in
allowing us to use much information collected during *The Sunday Times'*
own researches. In Gibraltar, David Connett, Richard Ellis, Jon Craig
and Jo Revill provided a vital verbatim account of the inquest. In
London, Max Prangnell and Andrew Hogg shared their own researches
with us, and Vera Taggart worked late into the night to produce a
pristine copy of the manuscript. In Belfast, Liam Clarke and his wife
Kathy provided us with a never-ending flow of objective research. This
book and its conclusions may not accurately reflect our colleagues' views
but they gave their time and advice nonetheless We owe them all.

Contents

Preface

On Sunday 6 March 1988, four SAS men opened fire on three IRA terrorists in a Gibraltar street and sparked perhaps the greatest controversy in the history of the Ulster troubles. When it was discovered that the terrorists had been unarmed, the SAS found itself on trial; the British government came under attack for supporting the Regiment without question, and the world was left in little doubt that the men from Hereford had committed a deliberate act of cold-blooded homicide, inflicting a surgical amputation of one of the IRA's most valued active service units.

In the political clamour, left and right took sides. The left screamed for a judicial inquiry and asked if the SAS was above the law. The right leapt to the defence of the Regiment, reminding critics of the atrocity the shootings had averted. For no one was in any doubt that the bomb the terrorists had planned to explode would have killed as many as a hundred military bandsmen, tourists, pensioners and schoolchildren in this British backwater that was physically far removed from the troubles of Ulster.

In the midst of the furore the media sought the answer to the crucial question: did the SAS open fire without mercy, regardless of the law and the terrorists' disputed attempts to surrender? It seemed the very essence of that peculiarly British regard for fair play was both on trial and on display. For where else in the world would the deaths of three such terrorists be so thoroughly questioned; where else in the world would a terrorist organization be given the air time and column inches to turn the killing of three of its members into a major propaganda coup?

Is fair play a luxury that cannot be afforded in the war

against terrorism, a luxury the British public must learn to dispense with if it wishes to win the war against the IRA?

Ambush sets out to answer those questions. It is a detailed investigation of the war at its most ruthless, the battle of nerves between the SAS and the IRA. It examines the roles and deployment of terrorist and SAS trooper alike; and it delves into the historic clashes between the two highly trained and highly motivated groups in search of clues that could explain why the three unarmed terrorists died in Gibraltar.

This book also reveals the knife edge on which the men from Hereford must balance. For those who wage war against terrorism cannot avoid overstepping the line between heroism and villainy. Already in this war they must risk their own lives and the lives of innocent bystanders. Can we really ask them, in that split second of confrontation, to risk these lives unnecessarily to give a terrorist bent on murder the benefit of the doubt for the sake of fair play?

JAMES ADAMS
ROBIN MORGAN
ANTHONY BAMBRIDGE

October 1988

PART ONE

CHAPTER 1

'Matters will be Resolved . . .'

The first clue was just a snippet, a fragment of information routinely collated and seized upon by a keen-eyed operative who realized its threatening significance. It had been picked up by electronic interception, the modern method of detection employed by the security forces in their continuing battle with the IRA. Army signals units operating in Northern Ireland play a vital role in providing a secure communications network for the military, for good communications are as important to a fighting force as the weapons they use. But the signals units perform another, equally crucial task. Using powerful receivers and a network of ground-based listening stations they regularly target known areas of IRA activity, sweeping the atmosphere for vital intelligence that might indicate the movement of known terrorists or reveal their planned bombings and assassinations.

Electronic intelligence, or ELINT, has become so sophisticated that the bug planted behind a wall-hanging, or secreted inside a telephone handset to pick up and relay a conversation in a room, has long been consigned to the dustbin of surveillance paraphernalia. Though micro-bugs still play a part in some operations, intelligence gathering in the eighties is a sophisticated science that owes more to the space age than to Guglielmo Marconi's invention of the wireless. Conversations can be eavesdropped on and recorded from long distances by microphones which can pick up and decode the vibrations on a window pane caused by voices in a room. It is a simple matter of pointing the microphone at the window from distances of up to a quarter of a mile and listening in through a pair of headphones. Other technology allows the signals units to listen in to telephone calls, now increasingly transmitted by microwave. It is a routine surveillance measure practised by the intelligence agencies of the world.

A computer constantly scans the microwave transmissions in search of key words; they might be the name of a terrorist, his telephone number, an address frequented by the IRA or even the name of a likely target. Whatever key word or sequence of numbers the computer has been programmed to pick up, once spoken during a telephone call, the key activates the computer's recording facility and the conversation is relayed to the computer screen of the duty intelligence analyst. It is here that the surveillance can go wrong; tired eyes or a momentary distraction can cause a clue to disappear into the memory banks where it may or may not be spotted at a later date. But there are safeguards and system checks that generally ensure vigilance so that, just as the CIA can listen in to the mobile phones of the Politburo's cars *en route* from their *dachas* to the Kremlin, the British security services too can electronically infiltrate the IRA. It was just such a key that gave the security forces access to an IRA operation from its very conception, an interception that was to give British intelligence a vital lead that would provide the SAS with the opportunity to register its second coup against the IRA in 1988.

It is rare that electronic surveillance provides such firm, unequivocal intelligence. Normally, collated material has to be sifted and assessed; for the IRA is wise to the methods available to the British security forces and has long observed the famous advice imparted in the sixties by the American senator, Barry Goldwater: 'If you don't want it known, don't use the phone.' This appears to have been taken to heart by the IRA. In the Sinn Fein offices in Falls Road, Belfast, there hangs a warning sign which says: 'This phone is bugged.' But not even the IRA can impose total security on its ranks and, inevitably, there is a degree of careless talk. The particular piece of evidence snatched from the air by the signals units' listening posts in June gave the first indication of the IRA operation to come. As always with nearly all intelligence, that fragment was put together with other information to produce a picture that was startlingly detailed and included the identity of the target and the names of some of those IRA men who would carry out the attack. It was a simple matter to relay the information to an excited joint intelligence committee, which quickly realized that the IRA active service unit identified by ELINT was operating in the same area where troops had recently discovered what could only be their personal weapons dump. The cache of guns was being kept under surveillance by camouflaged observers waiting for the IRA to collect their weapons.

In the middle of June, the snippet of information was passed on to Army intelligence in Ulster. The intelligence analysis assessed, correctly, that a retired UDR officer working in Omagh bandit country in the west of the tortured province had been selected as the latest target in the IRA's sustained campaign of assassinations which, the terrorist leaders hoped, would inexorably drain the lifeblood of commitment, loyalty and, not least, the recruiting capability of the Royal Ulster Constabulary and the Ulster Defence Regiment.

The security forces had numerous choices: they could warn the target and remove him to a place of safety, but he would still remain on the hit list. They could spring a trap on the terrorists as they collected their guns, but the Republicans would simply make a propaganda meal of it

5

with the usual explanation of 'innocent Irishmen stumbling on the guns by mistake'. The third option was by far the best: to catch the gunmen red-handed, not only in possession of the weapons but in the very act of using them. There was a problem however; detailed as it was – the intelligence gave them an indication of the target and the people involved – it did not reveal the other crucial component that was required to guarantee a successful ambush, namely *when*. The terrorists might choose to strike in the dead of night at the target's home, during daylight as he drove to the shops with his wife or as he left church. Twenty-four-hour surveillance would be costly and, if the active service unit decided to take their time, drawn-out. The longer protective surveillance continued, the greater the chance that the Active Service Unit or ASU or the target might slip the leash.

The only answer was an ambush. A trap would be laid and sprung when the terrorists had been drawn to a killing ground of the security forces' choosing. The chances of success in any counter-terrorist operation increase in direct proportion to the level of control that can be exercised over the action. Thus the side that can choose the time and the place to lay an irresistible bait has surprise on its side and the cards stacked heavily in its favour.

This then would be the scenario. The IRA active service unit was to be drawn unwittingly into a trap from which the only outcome would be surrender or death. Terrorists on active service are generally far too nervous and far too trigger-happy to throw their hands in the air and go meekly to an indeterminate prison sentence. When challenged they often respond with 'trigger fingers' in the hope of escape. Given the calibre of the IRA men involved, and the training of the men who would spring the trap, the likeliest outcome was death.

Like so many of the terrorists operating in Northern Ireland and in the Republic, the identities of the IRA men in the Omagh area were well known to members of the security forces. The two most prominent republicans in the area were the Harte brothers, Martin and Gerard, who had been identified as IRA gunmen by the security forces at least five years earlier. Gerard Harte, at twenty-nine, was the leader of the IRA unit in the area. He had drawn his twenty-five-year-old brother into the terrorist ranks in his teens. Both lived in Loughmacrory and were married with one child each. A third man, thirty-two-year-old Brian Mullen, who lived in Six Mile Cross, County Tyrone, completed the trio of terrorists who formed what the IRA grandly called the Mid-Tyrone Brigade. Though active terrorists they had been cunning enough to escape the clutches of the security forces on numerous occasions. But on army and police computers their names were linked to the murder of at least thirty-two men – including soldiers, policemen and former members of the security forces.

The security forces decided on a complicated operational plan. Members of the army's Intelligence and Security Group would keep the IRA arms dump under constant surveillance. The retired UDR man was informed of the plot on his life and agreed to establish a pattern of movement that the terrorists were certain to observe. He was supplied with a Leyland truck, which he drove to work every day along the B4, the Carrickmore to Omagh road. The cab was lined with armour plate to provide him with additional protection should the terrorists penetrate the security screen and attack without warning. The target's drive along a lonely country road took him through a strongly Republican area.

Everything was set, the arms cache was being watched, the routine that would attract the Harte brothers and their co-conspirator was established and the watchers settled down for what could be a long wait. If and when the guns were collected, the Leyland truck would inexplicably break down and await an attack from terrorists, who would be unaware the SAS were lying in wait around the immobilized vehicle.

In the event, the security services might have wished they had lured the active service unit out into the open much earlier. For they were about to be given a demonstration of the IRA's flexibility that would shock not only the British army but also Whitehall. In the midst of this surveillance operation, the IRA emphasized their terrorist skills and highlighted the chancy nature of intelligence gathering; for while planning the assassination of the former UDR man, the Harte brothers and the Mid-Tyrone IRA were planning a secondary action that would far surpass this in its horror.

At 12.30 on Saturday afternoon, 20 August 1988 a coach carrying thirty-five members of the 1st Battalion Light Infantry from Aldergrove Airport in Belfast to their barracks in Omagh travelled towards the Newtownsville road junction on the A5. This was the main road from Belfast and was used regularly by the army who made no attempt to disguise their vehicles or to vary their route to the barracks. It was a fatal mistake born of complacency, for such predictable behaviour presented the IRA with an easy target. Carefully hidden in a small trailer parked by the side of the road was 200 pounds of Semtex, the powerful Czech-made plastic explosive supplied by Colonel Gadaffi. Security forces believe that hiding in the long grass several hundred yards away was Gerard Harte, the leader of the Mid-Tyrone brigade. In his hand was a small ICOM walkie-talkie radio. Attached to it was a specially adapted encode/decode device that sent out a short electronic signal to the detonator in the explosives. As the coach with thirty-five relaxed soldiers on board, returning from leave, drew alongside, Harte thumbed the button and a fraction of a second later the trailer erupted in a roar. The blast picked up the coach and hurled it, shattered, through the air. It came to rest in a hedge on the other side of a ditch one hundred feet down the road. Where the trailer had stood there was now a crater in the road twelve feet wide and six feet deep. Harte and his accomplices lingered for only a second to survey the damage they had inflicted, before scurrying from their hiding place to a waiting car that would carry them out of the vicinity.

As the smoke cleared, eight soldiers lay dead and twenty-seven others injured, blown free of the wreckage into adjoining fields, or lying trapped in the tangled steel. It was a sickening scene of carnage that turned the stomachs of witnesses. Motorists who stopped to give first aid and comfort later described the devastation, with shattered limbs, separated from their bodies, lying amongst the charred wreck of the coach, of soldiers screaming in agony from blast burns and lacerations caused by flying glass. One soldier was seen crawling into a field as if to find cover. Ambulances ferried the injured to hospital and scenes-of-crime officers from the RUC began the gruesome duty of sifting the wreckage for clues.

It was a major IRA success. In the immediate aftermath of the explosion the republican propaganda machine celebrated the blood spilt in their cause and congratulated the active service unit responsible. In its exuberance the IRA identified the bombers as members of the Mid-Tyrone brigade. The men who had been expected to walk into an ambush on the road from Omagh to Carrickmore had caused a mass slaughter just a few miles south on the A5.

At first the army reacted with shock, turning in on itself, suspecting the IRA had an informer inside the armed forces who had leaked the timing of the bus journey. In reality however the IRA are poorly served by informers and instead rely almost entirely on painstaking surveillance in advance of their terrorist attacks. On this occasion the army played into their hands by using the same route regularly. The soldiers were hardly a difficult quarry to stalk.

After the attack there were calls from Loyalist politicians in Ulster and both Labour and Conservatives in London for government action to counter the IRA. More troops, internment and more 'surgical' methods were suggested for neutralizing the IRA. But the government expressed its revulsion and stood firm, refusing to act until it had the results of an urgently mounted security review. Meanwhile the RUC, with precious little hard evidence to rely on, did what any police force would do in the circumstances; it pulled in suspects for questioning. They knew the identity of all the IRA terrorists operating in the area and specifically the identities of the Mid-Tyrone brigade, and took in for questioning eight people they considered likely suspects. Among them was Martin Harte, who the security forces knew was planning the attack on the former UDR man. It was a futile effort, with little evidence, and faced with a terrorist well trained in the techniques of interrogation himself, they were forced to release Harte. Naturally he had an alibi already established before the button was pressed.

In the week following the bus bombing, Margaret Thatcher ordered a complete review of the security options for Northern Ireland. She had ordered a similar review in response to the murder of two Signals Regiment corporals in May, beaten and shot after being dragged from their car in broad daylight on a Belfast street. On that occasion the Northern Ireland Office led the investigation; and although it had promised a fundamental re-examination of security procedures it had done little more than produce a long-winded report that everybody ignored. This time, the prime minister was determined that action not words would emerge from the inquiries. On the Monday and Tuesday after the bombing of the soldiers' coach, there were a number of meetings between representatives from the intelligence services, the Home Office, the police and the military. When Tom King was driven the short distance from his office in Queen Anne's Gate to Downing Street on Wednesday he carried with him a fat list of options to present to the prime minister.

One option eagerly demanded by Loyalist politicians had long ago been ruled out as an effective counter-measure when the terrorists appeared to be gaining an upper hand. Sending more troops to Ulster had been dismissed ten years ago on the grounds that such a response only added weight to IRA propaganda claims that British troops were an army of occupation. While such action was a palliative for government critics in the short term, there were no long-term benefits. Besides, there was little to do that was not already being done by the RUC and the army; more troops would simply give the IRA more targets. This time, the measures that were agreed were of long-term value and were kept completely secret from the public. After the talks with Margaret Thatcher all Tom King would say was: 'Of the matters resolved, some will become apparent shortly in various particular ways.'

It had been agreed that there would be an increased use of undercover agents in Northern Ireland both to infiltrate the IRA and to gather more general information through covert observation. A major programme of new legislation had already been agreed to undercut the financial resources of the IRA, and this was to be given priority in the government's programme in the House of Commons. In addition, more SAS troops would be sent to the province to augment the intelligence and surveillance capabilities of the RUC and the army's existing units. Indeed, when Tom King left his meeting with the prime minister, the extra SAS were on standby at Hereford, awaiting the call that would dispatch them to RAF Lyneham, where a Hercules C130 transport plane stood ready on the apron to take them and their equipment to Ulster. As they flew out, the exultant Harte brothers, back in Tyrone, were planning their encore, an encore that they little realized would bring the curtain down on their campaign of terror.

On the night of Monday, 29 August, just nine days after the coach had been bombed, Tom King's promise that matters would be resolved drew close to fulfilment. The IRA gunmen who had jubilantly pressed the remote-control device that had devastated the lives of thirty-five servicemen approached, under cover of darkness, their secret arms cache. They were planning to carry out the attack on the retired UDR man and needed their rifles and pistols. They never knew they had been spotted by the British observation post which relayed the 'go' message back to base. In a meeting that night in Omagh barracks between the representatives from the security Service, the SAS and the army, it was unanimously agreed that an attack seemed likely the next morning. The IRA would be planning to lie in wait for the regular-as-clockwork Leyland truck, but the SAS had other ideas. Their own plan was put into operation immediately.

Early Tuesday morning an SAS man crept into the house of the former UDR officer, donned his overalls and drove off, leaving the target safely hidden at home. Shortly before dawn that morning, a three-man team from the SAS, dressed in plain clothes and armed with Browning 9mm pistols and Heckler and Koch MP5 automatic weapons, moved out from Omagh barracks and trekked by a looping circuitous route. Travelling east and skirting the B4 road, they covered five miles on foot across farm-land until they came to the pre-ordained killing ground, a deserted farm standing hard on the road outside the village of Drumnakilly. They immediately hid, one inside a derelict barn by the side of the road and two others in the hedgerows on either side of the road. The trap was ready.

It was a few hours later, with the sun already high in the sky, when the Leyland truck appeared around the bend in the road and began to reduce speed. An SAS man in hiding checked his watch; it was 9 a.m. The truck drew to a halt outside the deserted farmhouse and the driver jumped down from the cab, kicked a tyre and walked back to unholster the spare tyre and tools. As far as passing motorists were concerned, he was an unfortunate fellow with a flat tyre. The SAS man on board pretended to change the flat, taking the apparently damaged wheel off but not replacing it with the spare. Anyone who passed would imagine he was in the middle of changing the wheels, or had discovered damage to the spare and was awaiting a replacement. From here on the plan was straightforward; the IRA were planning an ambush and either they would come looking for the truck when it did show up, or word would reach them that it had broken down. Either way, its position would filter back to the Hartes and they would be drawn to a killing ground staked out by the SAS. It proved to be a long wait and the unfortunate driver soon ran out of components to check and double check under the lorry's bonnet as he waited for the ASU to get the message.

In fact he had to wait more than six hours. The men in the hedgerows and barn had been there even longer, but it was no hardship to men who stayed hidden from Argentinians for more than two weeks in the same hide during the Falklands War. At precisely 2.30 that afternoon four men wearing black balaclavas, blue boilersuits, and gloves burst into the McBride home on the south side of the B4. Inside the house were Justin, a local schoolteacher, his wife, their five children aged between seven and sixteen and two of their children's friends, Peter Kelly and Carl McAleer. The gang pushed the family into the kitchen and after ripping the telephone from the wall they asked Justin where the family car was. McBride said it was outside but pleaded with the gang: 'Please don't make me drive it.'

One of the gang reassured him and asked for the car keys. Another produced a bullet and placed it on the floor of the kitchen, warning the family, 'That's for you if you move out of this room.' The gang left in McBride's Fiat Regatta to conduct a swift reconnaissance of the area where the truck had broken down half a mile up the road. At 3.40 p.m. the gang appeared in the drive of The Brickhouse, a large house on the opposite side of the road to the McBrides'. The house is owned by Annie McAleer; she lives there with her son Pat who farms the surrounding fields. Also in the house were the milkman and Eamon McCullagh, a local insurance agent. Time was now short and unlike at the McBrides', the gang did not cut the telephone line; they even apologized for the trouble they were causing.

Three of the gang armed with AK47 assault rifles and a Webley .38 revolver piled into the McAleers' white Ford Sierra and sped off down the drive. The fourth member of the gang drove off in the McBrides' Fiat to wait at a pre-determined rendezvous where the terrorists could change cars immediately after the assassination to throw off any pursuit. It was a routine precaution straight out of the IRA's training manual, as was getting rid of their clothes. ASU members are instructed to discard and burn their gloves and overalls immediately after bombings or shootings because, in the past, several IRA men have been convicted on the strength of forensic evidence collected from clothing carrying traces of explosives or powder burns.

Like many such operations, the counter-terrorist effort depended on good and timely intelligence and the ambush was almost blown. As the IRA gang broke into the McBrides' house an innocent local boy, Paul Breen, spotted an army surveillance car in the driveway of a derelict house nearby. It had been keeping watch on the terrorists' movements and reporting back to base so that the SAS men lying in wait would know the enemy's every move. Realising they had been spotted and fearful of blowing the operation, the surveillance men sped off, leaving the SAS decoy and his protective cordon without a hint of the terrorists' movements after that. The first thing the men hidden at the derelict farm saw was a white Sierra bearing down on them. They had expected a Fiat Regatta from the McBrides' house. The decoy driver was desperately exposed; the nearest cover was the wall around the derelict farm buildings and that was ten yards away. Knowing the attack was imminent and sensing where the white Sierra was coming from, he sprinted over the ten-yard gap and immediately felt the thwack and whine of bullets around him. He recognized at once the distinctive staccato chatter of an AK assault rifle – it was the favoured weapon of the Warsaw Pact and the third world armies and terrorists groups it had supplied. As he ran for his life, he couldn't help noticing that the weapon was being fired automatically; the magazine was being emptied in his direction. Bullets sprayed around his feet and ricocheted off the walls of the nearby building. Miraculously, he was not hit and reached the safety of the brick gatepost.

His colleagues further down the road had been unable to pull their triggers during the initial burst of fire for fear of hitting the decoy. Once he had dived behind the brick pillar they had a clear view of the target. The Sierra drew to a screeching halt fifty yards past the truck and the three IRA prepared to jump out, run back and finish the UDR man off. As the car doors opened, the hidden SAS men opened up in a withering crossfire. All three IRA terrorists were killed instantly.

Inside the derelict barn the fourth SAS man had radioed base and called for a Lynx helicopter. Minutes after the shooting, it landed in a field nearby as army and RUC units sealed off the area. The four SAS men jumped aboard the helicopter which clattered back to Omagh barracks before press or public could arrive on the scene.

The shooting was considered a major triumph for the army, the Security Service and particularly for the SAS, who had clearly demonstrated the combination of patience and ruthlessness that has made them so feared by the IRA. But the response by the RUC and the Northern Ireland Office was a combination of bureaucratic fumbling and fear.

There is complete co-ordination of all covert operations in Northern Ireland between the RUC, the SAS and the different intelligence agencies. Indeed, the RUC often provide the very detailed intelligence that the SAS use for operations. So, the Omagh shooting came as no surprise but the standard police response on these occasions is to produce a bland statement that simply says the incident is being investigated. This was to prove inadequate.

On the political side, no authority is given by the Secretary of State for the use of the SAS in specific operations in the province. Indeed, it would be most unusual for ministers even to be informed about a forthcoming operation. Permission has been given for the deployment of the SAS in the province, and their use in specific operations is at the discretion of the Commander Land Forces. Outside Northern Ireland, and especially if the SAS are to be used abroad in an incident where shooting might occur, political authority is almost invariably sought.

Nobody wanted a repeat of the political row that arose from the Gibraltar shootings so the response of the publicity machine in Stormont and Whitehall was deliberately slow. Matters *had* been resolved but now they were playing into the hands of the IRA's propagandists.

The shooting happened just after 3.30 in the afternoon and word of the successful ambush reached London before 4.00 p.m. The security forces believe that the first to respond to the deaths was the IRA itself. A local priest was telephoned and alerted to the ambush and it was suggested he go down to the site of the shootings. Father

John Cargan was genuinely appalled at what he found. And in honestly voicing his horror the curate could hardly have served the terrorists better if the IRA had written his script themselves. His voice choking with emotion, Father Cargan reported to millions of television viewers that night: 'It is quite horrific. It is just the tragic circumstances in which people seem to die in this part of the world. Irrespective of who they are, this is not the way people should die. Two of them could possibly be identified but the third seemed to be fairly badly dismembered. There had been a hail of bullets.'

Sinn Fein, the political wing of the IRA, was not slow to capitalize: 'The three republicans killed by British soldiers this afternoon were courageous and committed young men who have paid a heavy price for peace in Ireland. Mrs Thatcher's government clearly ordered this bloodshed in a vain attempt to intimidate Irish people opposing British rule.'

To the consternation of the army, the spectre of a shoot-to-kill policy was being raised once again, and the British government remained silent in the face of this criticism. The lack of any official government explanation for the ambush served only to fuel press speculation that the Omagh ambush was the first evidence of the tough new government policy. In fact, it was nothing of the kind. Once again ill-informed judgements were being made in the absence of explanation; once again the SAS was being accused of murder. But there had been no change of policy; the growing conviction that the ambush at Drumnakilly was a revenge attack orchestrated by the government and the SAS was a nonsense that nobody seemed willing to explain. Inside the security forces there was exasperation; would no one tell the public that setting up such an ambush takes weeks of careful planning, intelligence-gathering and surveillance? Who would tell the public that the ambush was coincidental to the bombing of the Omagh-bound bus and not a consequence of it? Who would tell the public that the terrorists' arms dump had been under surveillance for weeks before the bombing, that their plan to assassinate the retired UDR man had resulted in the ambush project long before Harte pressed the button?

After the shooting, many parties interested in the Northern Ireland situation were tempted to draw their own conclusions. King's remarks had presaged a new British get-tough policy with 'shoot to kill' an essential ingredient. In Dublin, within an hour of the news of the shootings, the Irish prime minister, Charles Haughey, publicly asked the Anglo-Irish secretariat in Belfast for a full and urgent report. The following day the Irish government also asked for an early meeting of the Anglo-Irish ministerial conference to discuss the shooting and other security matters. Seamus Mallon, the deputy leader of the Social Democratic and Labour Party in Ulster, appeared on the Radio Four 'Today' programme asking if there was 'an alternative way of dealing with these men'?

'Were the soldiers,' he wondered, 'under orders simply to eliminate these people'. 'That,' he said, 'was not the way a proper civilized society should work.'

In London, Jim Marshall, Labour's Northern Ireland spokesman, said he would write to Tom King pressing for a full disclosure to ensure that there would be no repetition of the shoot-to-kill policy. His disquiet was later echoed by Labour's deputy leader, Roy Hattersley, who warned: 'There has to be an inquiry into the deaths . . . not carried out by people responsible for the killings.' Paddy Ashdown, leader of the Social and Liberal Democrats, and a former special forces combatant himself, added: 'Successful operations against terrorism can only take place where security forces operate and are seen to operate solely within the framework of the law.'

Publicly, the reaction of the Ministry of Defence was to

say nothing. Privately, officials were frustrated and dismayed. The air time given to the controversy had, in propaganda terms, turned the shooting into a Pyrrhic victory. It was a result, they believe, that could have been avoided. There are four basic factors, army sources contend, which would have convinced all but the most diehard conspiracy theorists and republicans that the killings were justified. The men who died were known terrorists, they were actively engaged on a murder mission, they were armed, and they fired first.

But the first police statement simply said that 'a shooting had occurred after soldiers encountered armed men in a vehicle. There are no security force casualties.' It neglected to mention the three dead terrorists, or the fact that the men had been on a murder mission, although it was revealed later that night that two rifles had been recovered from the scene.

It was then left to Sinn Fein to announce the names of the dead. But it was not until the following day that the RUC revealed that Gerard Harte was commanding officer of the Provisional IRA in Mid-Tyrone, and, in their eyes, a ruthless dedicated terrorist. The RUC refused to allow reporters to inspect the scene for themselves to draw up an accurate picture of what had happened. It was not until Friday afternoon that the cordon was lifted, and bullet holes from shots apparently fired by the terrorists could be seen clearly. By that weekend, concern in the Ministry of Defence and the intelligence services about how the IRA had turned yet another British triumph into a disaster had reached the ears of senior officials in the Northern Ireland Office. That week *The Sunday Times* in London published a detailed account of the mishandling of the information war and how the IRA had taken the initiative.

The prime minister finally decided to take action. She feels very strongly about the IRA and the way the media often panders to their propaganda machine. On this occasion the media wanted to tell the government story but were being frustrated in doing so. No. 10 Downing Street contacted Northern Ireland Secretary Tom King and told him to go public with a firm statement on the shooting which would back up the security forces. He argued that any debate on the actions of the SAS would interfere with the inquest into the killing of three IRA terrorists by the Regiment in Gibraltar. He was firmly overruled and that Sunday on the 'Weekend World' television programme, Tom King appeared to give the official version of events and to back up the SAS men who had sprung the trap.

Mr King said there was only one 'shoot to kill' policy currently in force and that was being carried out by the IRA. 'They go out and murder without scruples. Their policy is to shoot anybody, shoot first and shoot often.' He praised the courage and skill of the security forces and said that the three IRA men who were killed near Omagh had constituted 'a particularly nasty murder gang' with long criminal records behind them. There had been no doubt that they were out to kill and that the security forces had acted 'absolutely correctly' to prevent more killings. Mr King said he understood the IRA men had opened fire first in the exchanges that had ended in their deaths but there would be a full investigation to establish that the security forces had maintained the high standards of conduct expected of them.

It was too little, too late. Once again government ineptitude had generated sympathy for the republican cause; it appeared to many that the SAS, like the terrorists, were engaged in a private war in which both sides fired first, irrespective of the law. But the government seems oblivious to such criticism, for there is a flipside to the argument and one that is not lost on the military. If the IRA is convinced a shoot-to-kill policy is operating and expects its active service units to be wiped out whenever they come into contact with the SAS, then the creation of a few more martyrs to a cause that already has thousands is well worth the untidy disposal of hardened terrorists and the deterrent effect their deaths might have on other potential recruits.

The victory-turned-defeat at Omagh was just the latest battle in the long and bloody war. Today that war brings together two heavily armed, rigidly trained specialist élites – the most ruthless terrorist army in the world versus the best special force.

It was not always thus.

CHAPTER 2

The Sixty-Niners

At tea-time on 14 August 1969, as Protestant mobs went on the rampage in Belfast, the word went out. The Belfast brigade of the IRA called its men to arms. Catholics were being burnt out of their homes by drunken loyalists. The B specials, the exclusively Protestant reserve police force was either standing idly by or openly abetting the rioters. All that year the mercury had been rising. The civil rights marches, which in the early stages included both Catholics and Protestants, influenced by the rise of Martin Luther King's movement in America, demanded an end to the outrageous prejudice practised at all levels of Ulster society. At the beginning, the demands crossed the sectarian divide with calls for one man one vote and a fair housing allocations system. Protestants had, since partition, been favoured above Catholics in all walks; the best jobs in the administration, police and judiciary invariably went to Protestants, religion mattered more than qualifications or experience at job interviews, housing was allocated according to the colour of a family's politics; a loyalist Orangeman could always count on being considered first. The result of the marches, which were actually non-sectarian in that the cause drew support from both sides of the religious divide, was to instil paranoia in the staunchly loyalist cabals. They wrongly feared the civil rights movement was a cover for the rebirth of the IRA and the republican movement.

Temperatures rose, gangs inflamed by drink stopped people in the streets and asked their religion and beat up those who were brave enough to admit they were Catholics. A madness descended on the province. Catholic youths in Londonderry, mostly unemployed, began stoning police cars. Fearful of more street disturbances, civil rights marches were banned or re-routed, loyalist mobs that had lain in wait turned their attentions instead to Catholic homes and shops. By summer the situation was already out of hand, the largely peaceful civil rights movement had been hijacked by unemployed youths and left-wing radicals. Violence was met with violence, the hated B specials looked on or joined in and fear strengthened its grip on the province.

To this day there are loyalists and policemen who will swear the events of 1969 were fomented by the IRA and republicans yet the truth is as laughable as it is tragic. On 14 August, as the second in command of the Belfast brigade drove home from work, another riot was breaking out between Catholic and Protestant youths, homes were burning and the B specials, driving the Catholic mob back into its ghetto were being spurred on by loyalists. Drink, the Scarman report was to conclude later, played a large part in the violence, houses were being looted and fired.

The Belfast brigade that rallied to defend Catholic territory that night was hardly a force to be reckoned with. For the 'army' that both loyalists and politicians feared comprised no more than a dozen men in what was supposed to be its second largest stronghold outside Dublin. That night the Belfast 'brigade' managed to muster a few more volunteers. From long-abandoned caches in lofts, garden sheds and buried dustbins, it recovered its arsenal, hidden twenty years or more before, when the IRA had last assembled a force in the province. It was a woeful and rusty collection, a couple of Thompson sub-machine guns, a World War II .303 Lee Enfield rifle and half a dozen pistols badly in need of oil.

This then was the Irish Republican Army, the feared force of guerrilla fighters that threatened to overthrow British rule and return Ulster to its rightful owners in Dublin. That night three shots were fired from one of the pistols at a loyalist mob, headed by B specials, which was advancing on Catholic Divis Street. The insurrection, as the RUC saw it, had started. The inexorable slide into anarchy had begun and the police reacted with machine guns.

In August 1969 the IRA numbered no more than a couple of hundred men throughout Ireland. In Belfast there were perhaps fifty or so who considered themselves 'volunteers'; but half were old men, veterans of campaigns long forgotten in Whitehall, the other half were either social members who liked to listen to the veterans' reminiscences or young idealists who just wanted to belong. (One sixteen-year-old who had joined in 1966 was a boy called Gerry Adams who would rise to become the architect of the modern republican movement.) In 1969 it was only the old hands who could remember where guns might have been hidden years before, and few who had joined knew how to clean or load the arms they now unearthed. The IRA's orders in those early days had been simple: defend Catholic ground and fire only to warn off the mobs. Ill-equipped, badly led and undermanned, the IRA's skirmishing was a pathetic show of force; but the police reacted as if the Irish Army itself had been ordered in by Dublin. Armoured cars patrolled, their machine guns blazing, and many innocents were killed in the crossfire, including children asleep in their beds.

By the time Whitehall sent in the troops, at the RUC's request, the IRA, such as it was, was exhausted. Many had rushed to join its ranks but they were mainly young-sters desperate to get their hands on guns to fight back. When they realized there were no weapons to be had and the IRA volunteers found themselves unable to offer any-thing more than token shows of firepower (unleashing a few rounds of rare ammunition before retreating), the beleaguered Catholic minority turned on their self-styled defenders; the IRA became known as the I Ran Aways.

The Catholics believed that the IRA was no more than a small group of toy soldier romantics, albeit with a few brave men in their midst, and looked to British soldiers for help. The squaddies sent in by Whitehall were almost universally welcomed by the Catholic community as their protectors. The IRA retired to lick its wounds while British soldiers took what they believed to be their rightful place in the hearts of the Catholic minority. It was par-ticularly galling to the IRA to see the squaddies patrolling Catholic areas and receiving cheers and mugs of hot tea from the community. Such sights were commonplace and received wide attention from the world's media, rubbing yet more salt into the wounds of the republicans.

By now the IRA had split into two camps. The old-guard hardliners, disgusted with the movement's loss of credibility, had warned for years that the leadership's decision to run down the military activity (there had been no recruitment, arms training or drilling for years) had resulted in the current crisis. The leadership back in Dub-lin had long ago deserted the path of violence in favour of a socialist approach, the forlorn hope of uniting Catholic and Protestant working classes behind a republican ban-ner. The hardliners, certain that events would turn their way and that the current welcome of the British army would soon become outlived, plotted a coup.

Among those who were convinced that the IRA needed to return to truly republican and sectarian ideals and fight to expel British and loyalist rule, was Gerry Adams, then

25

nineteen years old, whose voice was being increasingly heard inside the still-small Belfast brigade.

As the new decade approached, the plotters cut off their links with the IRA's Dublin command and froze the few thousand pounds that Catholic business men had donated to its coffers. The breach came at a special IRA convention called in December 1969. The IRA leadership in Dublin still determined to press ahead with its long-held belief that a socialist alliance with the Protestant proletariat would best bring about a united Ireland. At the convention it was agreed that the IRA, through its political partner Sinn Fein, would recognize Dublin, Stormont and Whitehall and contest elections to the three houses. It was an extraordinary volte-face. That the three representative bodies which maintained the partition so bitterly opposed by the IRA were to be sanctioned by the leadership was too much for the hardliners to swallow.

One of the few men in the upper echelons who opposed the change in policy, Sean MacStiofain, raced back to Belfast to call a crisis summit of his supporters. The men in Dublin had lost their heads, he proclaimed, they were selling out to the Brits, they were no longer the 'real IRA'.

The Provisional IRA was born that night. The Belfast hardliners, those delegates from the south who had opposed the leadership and sided with MacStiofain and any others who had not agreed with their delegates' support of the policy change, reaffirmed their commitment to driving the British out of Ulster and reuniting Ireland.

For a time there were two IRAs active in Ireland, the Official IRA and the Provos. But MacStiofain had taken all the young bloods with him (except Martin McGuinness, the brilliant young Londonderry leader who stayed with the Officials for a while longer before cross-decking to the Provos). At times open warfare broke out among the opposing factions and more than once assassination squads reminiscent of the Mafia were sent out to execute the rival leaderships. The split tore families apart, with fathers and sons on opposing sides. But the Provos soon gained the upper hand.

It was perhaps inevitable that the Provos' desire to meet the loyalists and the Brits head on, while the Officials tried to talk, would eventually endear them to a Catholic population that had begun to tire of serving tea to soldiers in the streets. For the people who had welcomed the British army were quickly growing to hate it.

Predictably, as loyalist attacks continued and the British army found itself more and more a referee between two warring factions, the military started cracking heads on both sides. With each turn of the ratchet, resentment built. If a loyalist mob attempted to attack a Catholic march, the troops tried to keep them apart and were met with bricks and petrol bombs from both sides. Retaliating against young rioters eager to get at each other only angered the mothers, daughters and sisters who had to bandage the wounds. When tear gas drifted from the streets into homes and caused babies and grandparents to retch, Catholics turned on their protectors. It wasn't long before the Provos had began to re-establish themselves in the eyes of the minority as the defenders the Officials had failed to be in the summer of 1969. Recruitment soared, the ranks of the Provos swelled, the arsenal grew as pistols, air rifles and shotguns appeared in direct relation to the growing resentment against the army and police. Long before Bloody Sunday in 1972, when thirteen Catholics were shot dead by paratroopers who claimed they were returning fire that came from the ranks of a protest march, the Provos had re-established themselves as the real heroes of republicanism and the only effective answer to British rule.

When the Provos called their inaugural meeting in December 1969, they kept the structure that had survived the 1916 Easter Rising in Dublin. It was then that the IRA, with 1,500 volunteers, had become the official successor to the Irish Republican Brotherhood that had stood for a hundred years or more in opposition to British occupation. In five days the Rising had been crushed. Then, as in 1969, the British army had been hailed and the rebels had been jeered, but again, then, as in 1969, sentiment and support withered in the face of a heavy-handed British response. The brutality of the Black and Tans brought in to stamp out the remnants of the IRA was such that it served only to swell its ranks. By 1920 the Cork brigade of the IRA alone numbered more than 1,000 volunteers; partition and home rule for the south followed quickly.

In 1920 the IRA was undoubtedly a fighting force of sorts, an undercover army formed into company, battalion and brigade strength according to geographical locations. It numbered perhaps 100,000 men – though many were supporters uninvolved in fighting, providing intelligence, surveillance, money and valuable hideaways, for men on the run or arms. It fought a war of ambush and assassination, using its strength in depth to maintain a constant harassment over a wide area of Ireland rather than attempting another pitched battle as it had during the Easter Rising. On partition, which was largely supported by Irish citizens as it split the nation on sectarian lines, the IRA watched its support evaporate as a free Ireland emerged, and the republican army disappeared underground, outlawed and hunted by the new government in Dublin.

The IRA has never recaptured even a tenth of the support or sympathy it gained seventy years ago but it has stuck religiously to its rigid structure. To this day it is commanded by a supreme body, the Army Council, which comprises seven men. It is they who combine the role of directing the gunmen of the IRA and the political support groups of Sinn Fein, the propaganda wing of the IRA.

Often membership of both military and political wings is indistinguishable, with officers performing dual roles in both military and political wings.

At the head of the IRA is the Army Executive which is elected by the Army Convention, held only twice since 1969. The twelve-strong Executive elects the Army Council, which is the IRA's board of directors. In turn, the Council elects a chief of staff from amongst the seven-man leadership. He acts as both chairman, president and chief executive of the movement. Below the Army Council is the 'management team', a General Headquarters staff. GHQ is a twelve-man body tasked to put the council's decisions into operation. Below the GHQ are the IRA's Southern and Northern Commands which are split into Brigades, each covering a county or area of Ireland. The brigades are in turn sub-divided into battalions which themselves comprise companies formed by the grouping of Active Service Units, the 'troops at the sharp end'.

It is a self-deluding structure meant to give confidence to the membership and cause concern to the security forces. It bears little resemblance to the military structure it tries to emulate. If it did, the IRA might boast a membership of 100,000, similar to its zenith in 1920; but in reality the IRA of today boasts no more than thirty experienced gunmen and bombers, with perhaps twenty apprentices and up to 500 volunteers who can be called upon to support their operations.

The Active Service Units each have three or four members or, only on rare occasions, as many as eight or ten. They operate on a cell structure, controlled usually by only one man to limit the danger of betrayal. The ASUs tend to specialize in bombings, assassination, ambushes or mortar attacks and mostly operate outside the areas in which they live to confuse the security forces. They have enough autonomy to choose their own targets and it is up to their controller to decide if a planned action should be referred up the command structure to be vetted.

On paper, a number of ASUs form a company, several companies form a battalion and maybe two battalions comprise a brigade. In reality in outlying counties, one ASU, its controller and the volunteers who act as quartermasters, intelligence officers, lookouts and paymasters can be the entire staff of a brigade. Only in Belfast does the structure stand up to scrutiny, where the city's brigade is divided into four battalions, each operating one ASU. It is no secret to the security forces that the IRA has no more than a dozen Active Service Units in Ulster, a couple in the south and one operating in Europe; fifteen ASUs then, operating in fifteen 'battalion' areas in six 'brigades' (Belfast, Londonderry, Donegal, Tyrone/Monaghan, Armagh and Dublin).

The IRA maintains a semblance of democracy in that its constitution allows for a consecutive ladder of conventions right up to the top of its chain of command. A company convention can elect delegates to a battalion convention which in turn elects delegates to a brigade convention. That in turn elects representatives to a general army convention which elects the Army Council. In the 1920s it might have been a workable system but now, with so few operatives, the IRA would be constantly electing the same men to the next convention up the hierarchy. Only twice since the present troubles began has an army convention met, in 1969 when the result was the formation of the Provos, and in September 1986 in County Meath in the Republic when Gerry Adams mustered grass roots support against the hardliners who opposed his leadership.

The result is that the IRA remains a self-perpetuating oligarchy. The seven Army Council members also form the majority on the General Headquarters staff, making up the numbers with five high-ranking brigade commanders whom they themselves have appointed. The GHQ staff divide the IRA's operations between themselves, taking 'departmental responsibility' for 'engineering, publicity, operations, finance, intelligence, training,' etc. Each delegates similar duties to volunteers at brigade level and oversees the implementation of policy on the ground.

Such is the shortage of manpower that it is not unknown for a leading Sinn Fein official to be Army Councilman, a member of the General Headquarters staff, a brigade commander and an Active Service Unit leader. Equally a gunman can be a member of more than one active service unit, an expert bomb-builder who puts together the explosive components for the ASUs may himself be a specialist in assassinations or mortar attacks, the intelligence officer for an area may be an active service unit member in another.

The complex nature of the IRA's structure is deliberate, dictated by common sense and realism. Keeping the chain of command tightly in the grip of a few and the numbers of gunmen to a minimum reduces the risk of betrayal and clumsy or abortive actions due to over-enthusiasm or inexperience. It also recognizes the need to promote only those who are trained and dependable to the rank of active service. Security forces' success against the IRA have meant large numbers of IRA men killed or imprisoned. These successes have deterred many more from joining and the IRA's inability to train men in isolated areas of the south since Dublin clamped down on illegal training camps has further reduced the Army Council's ability to maintain a proficient stream of recruits. The days of the early seventies, when some were sent to Libya to be trained by Gadaffi's experts, have long gone – it is the IRA that is now considered the best terrorist group in the world, and expert enough to teach Gadaffi's training officers a few techniques.

Instead the IRA relies on the rehabilitation of terrorists freed from jail and the slow induction of promising talent to keep the active service units manned. The age of the gunmen mirrors the age of the movement's leadership; most of the ASU members have ten years or more experience and are now in their thirties and early forties, having joined the ranks as angry young men in 1969. It is this year from which they take their name – the Sixty-Niners – to distinguish themselves from the long-retired old guard and the yet-to-be-blooded new breed.

In the early seventies, British Army intelligence assessments considered the average IRA man to be a hard-drinking young lout who would have turned to petty crime and violence if the IRA hadn't offered him a cause. Much has changed since then. Many of these young firebrands have disappeared, have been killed planting bombs, shot trying to escape from a sniping operation, or jailed. Yet more have been weeded out by the IRA itself, ever vigilant for bad apples and always brutal in punishing them for 'anti-social behaviour' in the Catholic areas. What is left is a hardcore of volunteers who have matured and maintained their commitment. They have been educated by the intellectuals like Gerry Adams, and hardened by the determination of Whitehall not to give in to terrorism.

Most are employed, married men with children; Patrick Magee, the Brighton bomber, worked for *Republican News*, the official mouthpiece of Sinn Fein and the IRA; another IRA bomber, Patrick McNamee, worked in a factory making gaming machines (components, including tape and circuit boards, from the factory were later found in his bombs). All active service men get a £10 weekly allowance, double if they are married. It is an increasing financial burden to the IRA that the families of jailed IRA men also draw the allowance. The aim is that each active serviceman should immerse himself totally in the community so that he does not appear to be involved in terrorist activity. The truth is that the security forces know precisely the identities of both IRA leaders and gunmen but catching them in the act is the problem. They take elaborate precautions to escape surveillance and it is here that the vexed question, much debated in the eighties, of a shoot-to-kill policy can be most easily answered.

There is no doubt that when the SAS or the security forces catch IRA men in the act and open fire, they shoot to kill – it is the only guarantee that a terrorist can be immobilized and prevented from detonating a bomb or firing a weapon. The belief that someone can be winged and disarmed with a well-placed bullet is a Hollywood fantasy that fails to take account of human frailties such as nerves, fear and misjudgement. Terrorists with bullets in their brains have been known to carry on shooting. So if lives are considered to be threatened the only realistic reaction is to shoot to kill. Caught in the act of planting a bomb, or aiming a weapon, few terrorists can expect to live unless they can freeze in their tracks the moment a warning is barked. Critics of the security forces would argue that a deliberate shoot-to-kill policy is implemented when a terrorist presents himself as a legitimate target; in other words, caught in the act of aggression, a terrorist is killed and questions can be asked later, but the government can be relied on to back up the security forces.

The conspiracy theorists see black and white where there is rarely anything other than grey. If there were a deliberate shoot-to-kill policy there would be precious few gunmen operating in Ulster; as their identities are known it would be a simple matter to assassinate the ringleaders and their gunmen and any others who tried to take up arms. That this is not the case demonstrates that the rule of law governs the actions of the SAS, the Army and the RUC. In reality many IRA terrorists are arrested when caught in the act but many others are killed. The SAS say the records show they arrest seventy-five people for every twenty-five they kill. Undoubtedly there are many who set out to arrest terrorists and succeed. There are occasions when an SAS man or an RUC officer misjudges the reaction of a terrorist and opens fire rather than take any chances. Equally there have been occasions when individual servicemen or officers decided to open fire without warning, and there are other occasions where members of the security forces have called a challenge and been shot as a result. The question is not whether a shoot-to-kill policy is in operation – it obviously is not – but whether the authorities turn a blind eye or cover up on occasions when individual officers or servicemen kill without warning. To the men in the front line, such grey areas boil down to crude simplicities; the terrorist expects to be shot dead if caught in the act, threatening the lives of members of the security forces or innocent civilians. When a soldier opens fire on a terrrorist he fires not to wound (something only done in John Wayne films) but to kill. The security forces who live with the threat of death every day are bemused that anyone should care how an IRA gunman dies as long as one more mad dog is disposed of and the lives he might have planned to take are saved.

Within the infrastructure of the IRA the gunman's role is, like that of the RUC and the army that opposes it, desperately reliant upon logistical support. Just as the security forces need paymasters, clerks, medics, drivers, publicists, armourers and procurement executives to equip it, so too does the IRA. It is here that the manpower resources of the IRA are concentrated, and, as in the active service units, the GHQ staff and the Army Council, Sinn Fein members actively participate in fuelling the dual needs of the republican movement.

In recent years, the tandem strategy of the bullet and the ballot box has blurred in the public perception the division between the political and military infrastructures so that the two are almost indivisible. Funds that were once raised separately are now pooled and individual budgets decided by the Army Council. The lion's share goes to Sinn Fein to finance Gerry Adams' desire to see the party emerge as a political force. Since 1969, when the Belfast IRA's purse stretched to little more than £3,000 and the movement as a whole could muster little more than £20,000–30,000 in a year, Adams and his fellow Sixty-Niners have created an international underground corporation with profits of around £7 million and an annual turnover many times that figure.

In the early seventies active service units robbed banks and company payrolls to finance arms-buying abroad but quickly realized the disadvantages. In one celebrated case, the largely Catholic workforce of a factory robbed of its payroll marched on Sinn Fein offices demanding the return of their paypackets; they didn't see why the bairns should go hungry over the weekend and the menfolk without their drink so that the active service units could buy more guns and ammunition. Other funds were raised from IRA-controlled illegal drinking clubs that the terrorists licensed and ran in Catholic ghettos, from collection boxes passed round Irish communities as far apart as New York or Sydney, from protection rackets and simpler forms of extortion. But like the Mafia, the IRA was forced to find more sophisticated methods of raising funds – and hiding them from government agencies.

It hijacked and burnt out buses so that Belfast's transport system could not operate in Catholic areas and immediately set up its own black cab business, ferrying shoppers to the city centre for the same cost as a bus ticket and raising thousands of pounds every week from a monopoly maintained by violence. Within months the IRA had 600 cabs operating from two garages and had 'acquired' a part share in a tyre business, petrol stations and garages.

Smuggling was another revenue earner for the IRA which took advantage of the tax differentials between north and south to establish a thriving black market in consumer goods and livestock or grain that qualified for

35

EEC subsidies. It worked quite simply. At one time in the seventies Brussels paid pig farmers in the south up to £8 for each animal exported to the UK via Ulster. The same pig would simply travel north and, once its papers were stamped and the subsidy claimed, be smuggled back into the south along the labyrinth of country lanes and farm tracks to be returned north again. It was not unusual for the same animal or load of grain to repeat the process many times over.

Building contractors too found themselves paying a heavy price to keep the IRA from 'delaying' work on sites. It became normal practice for any contract awarded in the north to carry a ten per cent mark-up to the IRA or Protestant paramilitaries. Often the British government itself was paying the gunmen indirectly when it awarded contracts for public works such as new barracks, RUC stations or new government buildings, to builders who knew they would be out of business unless they co-operated with the Ulster Mafia.

Such scams were commonplace and netted hundreds of thousands of pounds for the IRA. It began to put its fingers in almost every till in the north; even the one-armed bandits in pubs and clubs were soon controlled by the two-armed kind, for one machine could net £27,000 in a good year. But, as the need for greater revenue-generating ventures increased with the IRA's expenditure, so too did the sophistication of the IRA's long term financial blueprint. The Mafia had seen the need to avoid probing government legislation by transferring its funds to legitimate businesses in America after World War II. In the eighties, Adams and the IRA also recognized the wisdom of this particular shift. They began buying into legitimate companies or setting up their own; and as the terrorists adapted to the society, some paramilitaries even began offering a security service generating hundreds of thousands of pounds. It was a perfect scheme; on the mainland uniformed guards patrolled building sites, warehouses and factories, but in Ulster they had no need of peaked caps, epaulettes, torches and keyrings, for the very knowledge that a front company for the IRA was maintaining security kept the petty thieves, burglars and vandals well away.

Adams moved the IRA funds into shops, estate agents and even craft markets selling tourist knick-knacks and may yet move it again since, in late 1988, the British government announced the formation of a new task force of inter-agency experts including policemen, customs officers, accountants and inland revenue investigators who are to be given sweeping powers to probe the books, bank accounts and business of those 'legitimate' IRA front companies.

Just why the IRA needed to increase its revenue – the organization itself has not grown significantly in numbers – is easily explained. As a simple terrorist organization carrying out sustained bombings and shootings, it needed only to keep its arms stocks high. But the sudden change in policy, in particular Adams' pursuit of political power, required vast expense. Sinn Fein set up advice centres and political officers, provided free legal advice, professional counselling and community help to endear itself to the isolated Catholic working class. It had to pay volunteers to police the areas and deal with the drug-traffickers, sexual offenders, burglars and vandals who dared commit crimes on the IRA's patch; it had to finance the growing publicity machine that is vital to any political cause. Increasing successes by the security forces in arresting volunteers required extra vigilance and therefore entailed more cost in hiding men and transferring them from one safe house to another. Arms caches unearthed by the security forces had to be replaced and weaponry updated. Technical research to make bombs more effective, costly operations taken to the mainland and Europe, visits abroad to argue the republican cause, the cost of inserting 'sleepers' in Britain,

Holland, Belgium and West Germany to await orders perhaps some years off; all were a huge drain on resources already stretched by the long-term strategy of buying into legitimate businesses.

The cost of fighting a handful of seats in the 1983 general election more than demonstrates the financial burden that Adams' strategy has put on the IRA. It cost Sinn Fein £137,000 to get Adams elected to a seat he refuses to take in the House of Commons. When the British, Irish and EEC elections, together with by-elections and local elections in districts and counties, are taken into account, the strain on IRA purse strings since it ended the policy of abstaining from electoral battles in 1981 has been huge.

Yet more cash has been required to update the IRA arsenals. In the eighties the security forces have scored considerable successes in disarming car bombs planted by the IRA and uncovering arms caches. The IRA has had to shop further afield for more compact weapons such as plastic explosives and Heckler and Koch sub-machine guns. Against this backdrop of huge electoral and military expenditure, the IRA lost a massive £2 million through court action in Dublin in 1985. In February that year, the Dáil passed new legislation that allowed it to seize that amount held in a Bank of Ireland bank account. The Dublin government believed the money had been earmarked to finance Sinn Fein's council election campaign in May that year. Its loss led to a bitter struggle within the IRA that almost saw Adams' demise.

Four hardline members of the IRA's ruling élite opposed Adams' ballot box strategy, arguing that it was contributing little to the removal of British rule in Ulster. Only a special army convention called by Adams and his supporters staved off defeat; and then only after he had prepared the ground by expelling four hardliners, including his one-time friend and fellow Belfast brigade commander Ivor Bell. The victory was a resounding endorsement of Adams' long-term strategy and increased his grip on the republican movement to the extent that the 1988 revenue of £7 million would be split 80%–20% in favour of political activity.

The breakdown of current IRA finances, which despite the shift towards legitimacy still rely on a bedrock of crime, is revealing. It is estimated that the IRA in 1987 raised the following:

- Robbery: The IRA staged armed robberies on 'unrepublican' targets in the south, netting £1.5m in 1987. In the north it stole only £30,000.
- Extortion brought in £500,000.
- Building site frauds including tax evasion raised £1.5m.
- Clubs and gaming machines earned £1 million (£250,000 from slot machines alone).
- Taxis brought in £175,000.
- Donations amounted to £100,000.
- Noraid, the North American support group for the IRA and Sinn Fein, sent £75,000.
- Moneylending at interest rates of between 35% and 40% raised £100,000.

In addition to this, the IRA raised more than £2 million from its 'legitimate' business interests which now include video rentals (of mostly pirated tapes), estate agencies and restaurants. The IRA has, in purely financial terms, become the most successful business in Ireland and would qualify, so the joke goes, for a Queen's Award for Industry.

From an estimated revenue of £7 million in 1987, the IRA's expenditure outside its vastly expensive political strategy includes £1 million for military operations, £62,000 in allowances to the active service units, and £125,000 a year to the dependents of republican prisoners. It is no surprise that the hardliners have fought to increase the size of their share in the IRA's financial success. They need money to finance the increasing costs of carrying their war away from the hard targets that have become too dangerous to attack. The SAS and the security forces have used their intelligence-gathering capabilities to good effect, laying well-planned ambushes to catch the IRA in the act of aggression. The switch to 'soft' targets in Britain and in Europe where the SAS do not routinely operate and where security is more relaxed was deliberately planned to avoid disasters like the ambush at Drumnakilly. It is the supreme irony that the IRA has turned full circle; for it was a policy decision taken by the terrorists in the mid-seventies which brought the SAS into the war. Then the IRA exported its violence to further its war against British rule and the SAS were sent in in response. Now the IRA is exporting violence to escape the consequences of 1975.

CHAPTER 3

Deployment

By December 1975, the IRA's campaign of terror had spread to the streets of London and the towns of the Home Counties. In four months of indiscriminate bombing the active service units sent to the British mainland had rocked the establishment on its heels and heightened security to a level not seen since World War II. Everyone was on alert, from policemen to nightwatchmen, shoppers eyed the corridors of stores and supermarkets for suspicious bags, posters began to appear warning London's population to be vigilant, publicans watched their customers suspiciously, searching for abnormal behaviour that might betray a terrorist's intent. The campaign had claimed the lives of nine people and injured 177.

The IRA had operated on the mainland before but never with such sustained skill, demonstrating good intelligence, and a competent network of supporters to carry weapons, provide safe houses and hide explosives. The apparent ease with which the IRA melted into British communities from which they launched their attacks and the speed with which they melted back into the landscape to escape detection, had the British police forces at full stretch.

The campaign had opened on 27 August with the bombing of the Caterham Arms public house in Surrey. The pub was a popular haunt of soldiers at a barracks fifty yards away and at the time of the bombing the bar was packed with a hundred customers. Thirty-three people were injured in the attack, eight of them seriously. Over the next few weeks, bombs went off at business premises, underground stations and restaurants. According to Seamus Twomey, then the IRA's chief of staff: 'by hitting restaurants, we were hitting the type of person that could bring pressure to bear on the British government.'

To keep the police on the defensive, the IRA also varied their attacks to hit specific individuals. One morning in October a bomb planted under the front wheel of a Jaguar car owned by the Conservative MP Hugh Fraser exploded without harming him after he delayed his drive to work to take a telephone call. The bomb killed instead Dr Gordon Hamilton Fairley, an eminent professor and leukemia specialist who was walking past with his dog. Ironically, if the IRA had killed their intended victim, they might well have suffered from the consequences. Fraser had been planning to drive Caroline Kennedy, the daughter of the later President Kennedy, to her art classes and the murder of a member of the famous and beloved American Catholic clan would hardly have endeared the republicans to Irish Americans.

In November, Ross McWhirter, the co-editor with his twin brother Norris of the *Guinness Book of Records*, was gunned down on the doorstep of his London home late at night after answering a ring at the doorbell. He had made no secret of his dislike for the IRA and had offered a £50,000 reward for information leading to the arrest of those responsible for the bombing campaign. The attack was notable for two reasons: it spelt out to the general public that the IRA would happily target civilians for execution if they felt they were hostile; and the active service unit was clearly flexible enough to attack freshly selected targets at random. Various other planned attacks, including the bombing of former prime minister

Edward Heath's home in London, were thwarted by police. After each bomb exploded, Scotland Yard's anti-terrorist squad made a detailed analysis of the methods and equipment used by the active service unit. They noticed that the terrorists generally struck on Sunday evenings, so it was decided to flood certain areas of London with 1,500 policemen for selected Sunday nights.

For several weeks Operation Combo drew a blank. But when for the first time the police switched the operation to a Saturday evening they got lucky. Around 9.30 on the evening of Saturday 6 December PC John Cook, who was patrolling in Mount Street, Mayfair, noticed a car with four men inside heading towards him from Park Lane. Two weeks earlier a bomb had gone off in Scotts fish restaurant in Mount Street, killing two diners and injuring many more. As a precaution two policemen had been stationed in a doorway opposite the restaurant although it was thought unlikely the terrorists would strike in the same place twice.

As the car drove past the restaurant, the stubby barrel of a machine gun appeared and raked the front of the building. As the car turned into Carlos Place beside the Connaught Hotel, the two policemen commandeered a taxi and gave chase, alerting other units in the area as they did so. The terrorists swiftly abandoned the car and in a running gun battle with police fled into Balcombe Street in Marylebone. In no. 22B, in their first floor council flat, John and Sheila Matthews were sitting watching 'Kojak' on television. Suddenly there was a knock at the door and John, a fifty-four-year-old postal worker, went to answer it. He was pushed back from the door by the four IRA gunmen who rushed in a group into the living room. John was tied up with a pair of tights the terrorists found conveniently lying on top of a pile of laundry and made to lie face down on a sofa.

Within half an hour the police had discovered where the

gunmen were hiding and the house was surrounded by armed men from the police Special Patrol Group.

A month before, two black men had taken control of the Spaghetti House restaurant in Knightsbridge and held staff and customers hostage; the police had successfully brought the situation to a peaceful conclusion. Although on this occasion they were determined to capture the terrorists, they were equally sure that concessions to the terrorists were out of the question. As the then police commissioner Sir Robert Mark said later: 'Though we were deeply concerned about the safety of the hostages, I did not consider for one moment that they were not expendable. I felt heartfelt sympathy for Mr and Mrs Matthews but felt that human life was of little importance when balanced against the principle that violence must not be allowed to succeed.'

This resolve was strengthened after police examined the abandoned car and the Mark 2 Sten gun they found inside it. Fingerprints in the car matched those of a man the police had codenamed Z, who was also believed responsible for the shooting of Ross McWhirter and a number of bombings. Police identified him as Harry Duggan, a twenty-four-year-old terrorist whose death the IRA had announced two years previously. He was supposed to have been killed when a bomb he was handling blew up in his face. He had been buried in an unmarked grave in the Republic near the border with the north. In fact, far from dying, Duggan had joined the other three members of the hit team and after a period of training had arrived in England to set up the terrorist campaign.

Negotiations with the terrorists were slow. For the first two days the hostages and the terrorists had no access to a lavatory and made use of empty drink bottles and towels. Then the terrorists accepted a chemical lavatory which was lowered to them by police in the flat above. But this willingness of the terrorists to negotiate was misleading and the stalemate continued for five days.

On the first day of the siege, two men from the SAS appeared at the scene at the request of the Home Office to offer their help. This was curtly rejected by the police who believed that the siege could be ended peacefully. They were also still basking in the generous publicity that had resulted from the Spaghetti House siege and had no intention of letting the Regiment, still hardly known of by the public at large, get any of the potential glory. By day five, the food and cigarettes inside the tiny flat were exhausted and the terrorists agreed to receive a supply of food. The police sent in sandwiches, coffee, soup and forty cigarettes. The terrorists had expected a full meal and in an irrational outburst of anger threw the food out of the window into the street.

By the seventh day, Friday, the terrorists had become desperate. They dressed Mrs Marshall in her overcoat and, armed with her car keys, prepared to make a break from the flat. In fact, the police had already located the car and let down the tyres. The terrorists' plan to escape was monitored through the bugs that had been planted in the walls of the flat and plans were laid to ambush the terrorists as they made their break. But these desperate measures, which would almost certainly have resulted in the death of Mrs Matthews, were never needed.

The previous day Robert Mark met the police negotiator, now himself Commissioner of Police, Peter Imbert, to discuss tactics. Although neither of the men wanted to call in the SAS with all the loss of face that implied, they ordered the Regiment to prepare a contingency plan for assaulting the flat. Meanwhile, they felt that the threat of their use might have some effect on the gunmen. So, the news was carefully leaked to both the *Daily Express* and to the BBC that the police were planning to call in the SAS. The following morning, just as the terrorists were about to leave the flat with Sheila Matthews, the BBC news broadcast that the government were considering sending in the SAS. As Sheila Matthews later recalled: 'That changed the gunmen's minds. It was after that news that they changed their minds about the breakout.'

At 1.45 p.m. the terrorists telephoned the police and began the discussions that led to their surrender. As Sir Robert Mark put it: 'They could hardly surrender fast enough. They went through the pretence of delay, demanding a meal as a show of bravado, but not for them the prospect of dying for Mother Ireland. They came out shaking like leaves exposed in their true colours for all the world to see.'

It was an astonishing turnaround, and to the waiting police it was clear that the mere mention of the possibility of the SAS arriving on the scene had been sufficient to break the terrorists' will. Shortly afterwards, a hot meal was lowered to the flat and Mrs Matthews was released. The successful end to the siege brought to a close one of the bloodiest and most successful years in the IRA's history. It also clearly demonstrated to the police, the army and, most importantly, to the British government, the effect the SAS had on hardened IRA terrorists.

Although this was the most visible sign of the influence of the Regiment on the war with the IRA, there were other circumstances that year that drew together the IRA and the SAS on what was to be an inevitable collision course. For the Special Air Service, the 1970s heralded a period of great change. The Regiment had proved its worth in World War II in a series of campaigns in Africa and Europe. But, in the immediate aftermath of the war, neither politicians nor senior army officers saw the need for a unit of irregulars such as the SAS. There had always been a distrust of such special forces which operated outside the regular service structure using innovative tactics that were rarely written down in formal training manuals. The result of such a short term attitude was the disbanding of the SAS as a regular unit and the formation of 21 SAS (a Territorial Army reserve force) in 1945.

But such complacency soon became apparent. Peace in Europe and the Pacific had left many disparate and well-armed revolutionary groups eager for power. Beginning with Palestine a rash of small wars broke out around the world. These were in large part conflicts created by countries fighting for their independence from colonial control. All empires suffered; French, Belgian and British colonial power was being fired at across the globe. This was a new kind of warfare for which the colonial powers were ill-equipped. In the past such spasms were swiftly dealt with by the application of overwhelming conventional forces. But the opponents had become more sophisticated, using the tactics of the guerrilla – clandestine operations by a secret army with no visible logistical support – to prosecute their war.

The first of these wars took place in Malaya, where 10,000 communist guerrillas, supported from neighbouring Indonesia, waged a campaign of sabotage and assassination from their bases hidden in the dense jungle. The British government was unwilling to commit the numbers of men necessary to fight and win a conventional war against the guerrillas so soon after the end of World War II, and their early efforts using regular troops in small numbers were not successful. The guerrillas who knew the terrain had the advantage of a safe haven across the border and received intelligence and supplies from local people; they proved impossible to find.

Then in 1950 the Director of Operations for the campaign consulted a former SAS officer, Mike Calvert, to ask his views. He produced an assessment of the war that twenty-five years later would conveniently transfer from the Malayan jungle to the farmland of South Armagh.

His assessment essentially argued that the British had failed to understand the nature of the war and a new kind of unconventional approach was needed if the British were to have any hope of containing the problem. He made three key points:

1. A specially trained force should be created that would be able to fight the communists on their own territory. This force would be able to live for long periods in the jungle and would not only gather valuable intelligence but would also deny the guerrillas any safe haven.

2. Living and operating from their jungle hideouts, the guerrillas were almost entirely dependent on the rural population for food, money and intelligence. Calvert saw that if population areas could be denied them, then the guerrillas would be forced out into the open where they would be vulnerable.

3. If the guerrillas could force the population to support them through terrorism, Calvert felt that the counter force could win them over using a more humanitarian approach. This was the beginning of what became known as a hearts-and-minds campaign and has been a cornerstone of all counter-insurgency operations ever since.

Calvert's recommendations were accepted and the SAS was re-formed as part of the regular army. They went on to fight successfully in Malaya and then were used to fight other brush fires in Oman, Brunei, Borneo and Aden. In 1957 they were formally established as 22 SAS Regiment based at Hereford, where they remain today. But by the mid-1970s their counter-insurgency role was changing. The British empire had virtually disappeared. The last of the major conflicts involving the SAS, the war in Oman against guerrillas supported by South Yemen, was almost over, with the guerrillas defeated and stability restored. While guerrilla warfare remained popular, it was not generally aimed at Britain and so the SAS were no longer deemed essential to the nation's defence. Once again the SAS feared that they might be relegated to Territorial status unless they carved out a new role for themselves.

Three events conspired together to help them in their search for a different identity. First, the Security Service and the Secret Intelligence Service (more commonly but incorrectly known as MI5 and MI6) had rediscovered a need for an 'active arm'. Since the ending of the Cold War and the arrival of high technology intelligence systems such as satellites, it was felt there was no longer any real need for the James Bond type of secret agent. But it was now recognized that there was a real need for the capable agent on the ground, the highly trained individual who could go into a foreign country to check out information or the team who could successfully extract a defector from a hostile country. The intelligence community lacked those skills and were anxious to redevelop them. Second, terrorism had erupted across the world in 1968. This was a new phenomenon that took advantage of all the modern world's vulnerabilities: modern communications that brought a single bombing or killing into a million households; an arms network that allowed terrorists easy access to weapons; a worldwide intelligence system comprising agencies that hardly spoke to each other and whose collection systems were aimed almost entirely at the communist bloc; and police forces who were untrained to deal with bombing and kidnappings. It was a time when a handful of killers could attack a nation's soft underbelly and rip it apart.

Finally, there was Northern Ireland. After more than forty years of relative inactivity, the IRA emerged once again in the 1960s at a time when revolution was spreading and terrorism in the Middle East, Europe and the Americas was not only fashionable but looked like being successful. Of course, there had always been discrimination against Catholics by Protestants in Northern Ireland where the Catholic unemployment rate was three times that of the Protestant working classes and all the good housing seemed to go to the Protestants. Into this fertile ground of social oppression, where even the police were seen as tools of Protestant oppression, a new, militant IRA found plentiful support. The split between the traditional Official IRA and the more violent Provisionals had severed many of the links with the old guard who favoured change through political action; the way had been left open for the Provisional IRA to rebuild the republican army and launch a new and violent campaign to establish a united Ireland.

The violence had rapidly escalated. In 1969 there were only eight terrorist incidents, all of them involving bombs. A year later there were 383 incidents, including 213 shootings and 170 bombings. By 1972 there were 12,481 incidents, including 10,628 shootings and 1,853 bombings. The government response to the deteriorating security situation was to send in the army to gain control of the streets and the countryside with their massive presence. This had some effect on reducing the number of attacks but they were still at an unacceptable level. There was a stand-off in which the IRA and the army traded blows but neither had an edge that might ensure defeat of the other.

Northern Ireland was the perfect opportunity for the SAS to demonstrate their skills. It was the type of counter-insurgency war of which they had gained considerable experience. While they might not have the freedom to operate as ruthlessly as they had in Oman or Malaya, where few people cared about the circumstances in which unnamed jungle revolutionaries died, they were still better equipped than any regular army unit to take on the IRA at their own version of the undercover war. Colonel John Waddy of 22 SAS drafted what he hoped would be a blueprint that would provide a coherent role for the SAS and secure its future. In addition to its normal wartime tasks of surveillance and sabotage deep behind enemy lines, Waddy proposed that the Regiment would have a recognized role in countering terrorism both in gathering intelligence which would allow the government to pre-empt planned terrorist acts and, if that failed, to respond to specific acts. In addition, the Regiment would be the active arm of the intelligence services going into hostile territory on joint operations with both the SS and SIS.

While the SAS were developing a new role and eagerly looking for an opportunity to show their mettle, the Provisional IRA, too, had moved on from their first wave of terrorism. At the beginning of 1975, the government and the IRA had agreed on a ceasefire. Both sides needed a breathing space; the government because it hoped that political initiatives involving all the different parties might bear fruit; and the IRA because it had been losing support among the Catholics as the violence of the campaign escalated. The IRA saw other advantages to the truce as well. Under the agreement, they were allowed to establish centres to monitor the working of the ceasefire and the government also agreed to allow a number of suspected terrorists who had been interned to be released. The IRA were therefore able to take back into their ranks some experienced gunmen while at the same time working in the community as quasi-policemen with official government sanction.

The security forces were furious at the deal struck by the government. In 1974 they had cut the number of terrorist incidents by a third and the army assessment was that for the first time since the violence began five years before they had the IRA on the run. The commander in Northern Ireland, Sir Frank King, argued strongly that given a bit more time the terrorists would have been defeated. This demonstrated a misunderstanding of the enduring nature of terrorism but even so there is no doubt that the ceasefire gave the IRA an opportunity to regroup, which they used to good advantage.

While the government and the IRA may have agreed a truce, no such agreement had been reached between the Provisionals and the Protestant paramilitaries. Overall, the level of attacks fell dramatically from 3,206 in 1974 to 1,805 in 1975. But some of the most savage attacks of the whole war occurred during the ceasefire. In April seven people were killed and seventy hurt in two Belfast pub bombings. The situation deteriorated so that every week there was an attack by the IRA on a Protestant target which was immediately followed by Protestant retaliation. The tit-for-tat killings spread to the rural areas. In a particularly horrible crime, seven men wearing the uniform of the Ulster Defence Regiment stopped a minibus carrying members of the popular pop group Miami Showband back to Dublin after playing at a concert in the north. The attackers exploded a bomb that killed three members of the group and two of the terrorists.

The attack caused outrage in north and south and across the sectarian divide. There was little doubt the IRA would retaliate and they did so the following month by blowing up a Belfast bar and killing five people. The conflict then degenerated into a series of increasingly violent attacks on civilians, both Catholic and Protestant, which created an appalling climate of fear in the community. At the same time, the IRA were carrying out their terrorist attacks in Britain. Almost unbelievably the British government maintained the pretence of the ceasefire, although the pressure from the army, the intelligence community and politicians on the Labour Home Secretary, Merlyn Rees, was becoming intolerable. In November he ordered the IRA advice centres to be closed and the truce came to an end.

On 5 January, six armed men flagged down a bus near the South Armagh village of Whitecross. The twelve people on board were ordered out and asked their religion. The Catholic driver was allowed to go and the remainder were machine gunned as they stood by the vehicle. Ten of the men were killed instantly and the eleventh seriously injured. Although the IRA argued that the ambush was simply in retaliation for Protestant attacks in the area, the sheer cold-blooded savagery appalled the Irish community and finally provoked the British government to take action.

But deciding what type of action was to be taken caused the government some difficulties. There were already 15,000 troops operating in the province and an increase in their numbers was not practical given the army's commitments to Nato. In any event, the general public would hardly be reassured by a modest increase in regular troops who were clearly failing to control the violence. Exactly who in the government thought of sending the SAS is not known, but Prime Minister Harold Wilson, who knew very little about the Regiment, seized on the idea.

He had been impressed with the successful resolution of the Balcombe Street siege where the mere threat of the SAS had been enough to force hardened IRA terrorists to surrender. Also, Wilson loved the secret world of intelligence and covert action and found the idea attractive both personally and in public relations terms. He could shroud the details of the plan behind the blanket cover of 'national security'.

On 7 January Wilson ordered the Regiment to send troops to Northern Ireland. Their operations would be restricted to South Armagh. The decision was announced in the House of Commons the next day. The exact number of troops to be sent was kept a closely guarded secret. That day Wilson appeared on television to explain his decision: 'We have a very special situation in Armagh. It is not merely that it is along the border and that people are doing these dreadful atrocities and then hopping back. But we were in danger of seeing a kind of gun-law regime develop on both sides. And we felt it right to put in the SAS, who are highly trained, skilled and courageous.' He added that 'some of the descriptions of the SAS were a bit like science fiction.' Merlyn Rees, the then Northern Ireland Secretary, put the decision into context. It was, he said 'more presentational and mystique-making than anything else.'

Members of the SAS tell a tale, which may well be apocryphal, that after Wilson ordered the SAS in he asked a senior civil servant how many would be sent. He replied, 'Five'. Wilson then wrote down '500'. 'No, Prime Minister,' protested the official, 'Five.' In fact Wilson did not believe that so few could really make any difference to the conflict. But as there were only some 400 members of the SAS at that time, there was no prospect of them being sent in large numbers.

In the event, eleven members of the Regiment flew to Belfast that week and moved down to Armagh to carry out a detailed intelligence assessment and reconnaissance. By April, sixty men, a full squadron plus logistics support, were stationed in the province. These were days before the SAS had been catapulted on to the world stage, to be caught in the glare of television cameras in the very act of counter espionage, insurgency and terrorism. In 1975 the IRA feared the SAS' reputation. They could not know that just as they themselves had sprung phoenix-like from the ashes of 1969 and the I Ran Away debacle, so too had the SAS been close to extinction before events turned in its favour in the early seventies. Like the general public outside Hereford, Whitehall and the defence establishment, the IRA had never been given the opportunity to peer behind the veil of secrecy and mystique that had shrouded the SAS from prying eyes. Had they done so the IRA would have concluded that their ranks had little to fear from the SAS in 1969, but much more to fear in 1976.

CHAPTER 4

The Regiment

It took nearly forty years for the veil to be lifted. On an afternoon in London in 1980, black clad members of the Regiment scaled the walls of the Iranian embassy in Princes Gate and killed all but one of the terrorist gang holding twenty hostages inside. In eleven minutes relayed to television screens across the world by the picket of international camera crews that had gathered in the early days of the siege, the SAS emerged from its self-imposed shadow to spawn a hundred books and thousands of newspaper articles on its history and its talents. The men whom the IRA had feared long before Harold Wilson deployed them in 1976 had ably demonstrated why the terrorists should beware.

When the SAS were formed in 1941 to fight behind German lines the founder and commander of the Regiment, David Stirling, determined that the ideal recruit would have to satisfy four key requirements. Stirling, a Cambridge-educated Scots Guards officer who had transferred to one of the first commando units in 1940, had pencilled a proposal from his hospital bed in Cairo to the Commander in Chief of Allied Forces in the Middle East, General Claude Auchinleck. But in an episode that has entered military legend, Stirling decided the message from a mere subaltern would never reach the in-tray of a general;

he took up his crutches, climbed over the fence of allied command headquarters in Cairo and pressed his point in person. Soon he was a captain in charge of the Special Air Service, a motley collection of sixty or so fierce commandos whose job it would be to strike deep into the heart of enemy territory and shatter supply lines and dumps at will.

While the role of the SAS has expanded massively since then, the raw recruit must still satisfy the same requirements laid down by Stirling nearly fifty years ago and since explained in more detail. The SAS soldier must:

1. Engage in the unrelenting pursuit of excellence.

2. Maintain the highest standards of discipline in all aspects of daily life, from the occasional precision drilling on the parade ground to his personal turnout on leave. 'We always reckoned that a high standard of self discipline in each soldier was the only effective foundation for Regimental discipline. Commitment to the SAS pursuit of excellence becomes a sham if any single one of the disciplinary standards is allowed to slip.'

3. Brook no sense of class, and particularly not among wives. 'This might sound a bit portentous but it epitomizes the SAS philosophy. The traditional idea of a crack regiment was one offered by the aristocracy and, indeed, these regiments deservedly won great renown for their dependability and their gallantry in wartime and for their parade ground panache in peacetime. In the SAS we share with the Brigade of Guards a deep respect for quality, but we have an entirely different outlook. We believe, as did the ancient Greeks who originated the word aristocracy, that every man with the right attitudes and talents, regardless of birth and riches, has a capacity in his own lifetime of reaching that status in its true sense; in fact in our SAS context an individual soldier might prefer to go on serving as an NCO rather than have to leave the Regiment in order to obtain an officer's commission. All ranks in the SAS are of one company in which a sense of class is both alien and ludicrous. A visit to the sergeants' mess at SAS HQ in Hereford vividly conveys what I mean.'

4. Display humility and humour. 'Both these virtues are indispensable in the everyday life of officers and men – particularly so in the case of the SAS which is often regarded as an élite regiment. Without frequent recourse to humour and humility, our special status could cause resentment in other units of the British army and an unbecoming conceit and big-headedness in our own soldiers.'

These simple principles persisted throughout the SAS's history, during periods of action and inaction. Between 1969 and 1972 the SAS was totally committed elsewhere. The Oman war was beginning to make heavy demands on the Regiment and no one could be spared for Northern Ireland.

But, as always, the Regiment was looking ahead. There was a recognition that with the end of the British empire, the era of the small war might be ending and new roles would need to be found if the Regiment was to survive untouched.

The perfect oppportunty presented itself on 5 September, 1972, when eight men belonging to the Palestinian terrorist group Black September scaled the six-foot-high wire fence that surrounded the Olympic village in Munich. They burst into a room housing eleven Israeli athletes, shot two of them and held the others hostage. The terrorists demanded the release of PLO prisoners held in Israel and members of the German Red Army Faction held in West Germany. They also demanded a plane to fly them to Cairo.

The West German government was totally unprepared for such an attack. It had no specially trained counterterrorist forces and after seventeen hours of negotiations appeared to concede to the terrorists' demands. They were flown in two helicopters to Furstenfeldbruk military airbase and as they prepared to fly on to catch a civilian flight, hidden snipers opened fire. Visibility was poor, the range too long and two of the snipers lost their nerve. The result was that the terrorists had time to blow up both helicopters and nine Israeli athletes lost their lives. It was a military disaster that could have been avoided.

The lesson was not lost on the British. In order to avoid such a fiasco on their own soil, the government turned to the SAS. The IRA may have looked upon it for many years as a counter-revolutionary force but it was only now that it grabbed that role for itself. Authority was given to establish a Counter Revolutionary Warfare wing (CRW) within the regiment. It would have a permanent staff of around twenty who would be responsible for training every member of the SAS Regiment's squadrons in counter-terrorism techniques. At any time one squadron would be responsible for the counter-terrorism role, each taking it in turns to be on twenty-four-hour standby. The duty squadrons were divided into troops, known as a SP (Special Projects) Team for no one could guarantee terrorists might not strike twice, hijacking an aircraft at Heathrow while holding hostages in a bank in London. That decision was the turning point that saw the SAS grow again, gaining momentum in the bandit country of South Armagh, winning prestige in a single action in London in 1980 and, inevitably, drawing flak for its singularly ruthless reputation. To understand its role in Northern Ireland it is necessary to understand the formation, motivation, training and techniques of the SAS; for it grew quickly, from a special force to be deployed in time of war, to a political tool that could be wielded like a surgeon's knife on a cancer.

Today the SAS is divided into four 'Sabre', or combat, squadrons bearing the initials A, B, D and G. Each squadron is broken down into four operational troops of fifteen men and one officer. Each troop must be expert in either sea warfare, mountain warfare, parachuting or other forms of airborne assault such as hang gliding and mobility (using specially converted Range Rovers and Land Rovers). Each troop is sub-divided into four-man working units. It is the four-man unit that is generally used in the field. While each member of the unit is required to be generally proficient in such skills as navigation and weapons handling, every member of a unit has to have a speciality of communications, medicine, transport or explosives. There is a great deal of cross training so that every person can do someone else's job in an emergency.

Recruits to the Regiment are drawn from all over the British army and have to go through an exceptionally tough selection course which places a heavy emphasis on stamina, initiative, intelligence and leadership. The initial selection course is based at SAS headquarters at Sterling Lines, Hereford, named after the Regiment's founder, and lasts around a month. It includes intelligence tests with a particular emphasis on hand-eye co-ordination and a gruelling timed forty-five-mile-route march in full kit over the Brecon Beacons in Wales. All recruits are tested for their ability to resist interrogation. It is recognized that modern drugs make any successful resistance over a long period unlikely, so the training is designed to make sure all SAS men can hold out long enough for the mission to be carried out or for a rescue to be affected. These interrogation sessions were exceptionally tough, as one intelligence officer quoted in John Strawson's *History of the SAS* explained:

> 'It is quite amazing, off the record, how many of our big, tough soldiers in fact weren't quite so tough under the skin. No physical violence whatsoever, it was purely mental. For example, you'd sit a chap down in a chair and you'd put a latrine bucket over his head, and a chap would sit there going bong, bong, bong with a spoon for about half an hour, and then he'd be taken outside and tied to a tree that had been smeared with jam, with ankle cuffs and wrist cuffs, so he had to keep away

from the tree, because the ants were pouring up and down it, red ants, all sorts of lovely tortures they devised these chaps. If you take a bicycle wheel with the spindle still in the centre, and put it on the ground and make you kneel inside the rim on spokes, with your hands behind your back tied to a rope over the ceiling, and you had to keep this bicycle wheel absolutely dead flat. You can't do it because you get cramp in the ankles and knees and every time you moved, you got a jerk on the wrists and it was very painful. It didn't leave any mark or sign; you weren't actually beaten up . . . It was great fun and it gave everybody a very good insight into the sort of chaps you had. You picked out then the really good and you picked out the doubtful ones.'

It also appears that a particularly malevolent SAS instructor had been seeing too many Buster Keaton movies. There was a convenient railway line at Hereford with a small siding. An SAS man would be hooded, taken to the railway track and tied down with his neck on the rail. His colleagues would stand around chatting and then be apparently surprised at the sound of a train approaching. Cursing and with a note of increasing alarm they would try and fail to untie the victim. As the train approached, the victim would be certain he was going to die, only to find that he was tied to the unused side track and the train had passed on the main line. Another tactic involved putting a man's head into a bag of snakes, which, while harmless, brought to life a deep-seated phobia present in many people.

Such sadistic training methods were of questionable value and were stopped, in part following a complaint from a general's son who had suffered.

The brutality of such interrogation techniques has now been abandoned but still recruits are expected to undergo at least thirty-six hours of continuous interrogation when, although there is no physical violence, they will be deprived of sleep and light to disorientate them.

Officer selection is made twice a year when an average of three men are chosen for a three-year secondment to the

Regiment. Other ranks can stay in the SAS as long as they pass the annual performance assessment. If recruits pass the first hurdle after a month then they move on to a more specialist six-month training course attached to a squadron.

It is a fundamental tenet of the SAS that every member of the Regiment is continually being examined for competence, and the learning process is also continuous. Any member of the SAS will always claim, quite truthfully, that he is being continuously trained.

The development of a counter-terrorist role led to a number of changes at Hereford. In particular, a building known as The Killing House was constructed to train Special Projects teams in hostage rescue and rapid entry into defended buildings. On the course, known as Close Quarter Battle or CQB, each soldier is taught entry into the building from ground level and from the roof. Inside the building, a single room was set up to imitate a room where hostages might be held. Inside the room, live 'hostages' – who were in reality SAS men – would be mixed in random patterns with 'terrorist' dummies. Frequently the room would be in darkness. The SAS rescue team would be expected to burst into the room and, in under four seconds, distinguish between the hostages and the terrorists and shoot the terrorists. Live ammunition is used by the rescuers so the stresses are similar to those in real life. As the scenario would be changed regularly by the instructors the rescue team had to develop very fast and accurate reactions to avoid killing the colleagues acting as hostages. This proved a very effective training method until nearly three years ago when an SAS sergeant acting as a hostage in the room moved at the wrong moment and was killed instantly by a shot to the head.

Now, a new kind of killing room has been designed that requires the same responses but, through the use of film projectors, avoids the dangers of SAS men being killed, by keeping terrorists and hostages in one room and the rescue team in another. Two rooms are constructed side by side inside the Killing House.

Above the door in the terrorist room is a camera which records every movement and relays it to the room next door. In that room is a door and above it a projector relaying on to the three opposite walls the image of the terrorists and hostages. The terrorists have a duplicate image of any movement that takes place in the room next to them. The SAS men burst in, see in front of them the image of the room next door and fire their weapons at the terrorists who are responding to what they see on the projector in their room.

As the action is over in around four seconds, there is no time for anyone to discover that the action is on a screen and not real. The system has the added advantage that every shot can be recorded on film and its impact can be measured on the screen. When the videos are played back, both sides can measure their response times and accuracy.

Special materials are used on the walls that make them sound-proof and bullet-proof. The average man going through a training week at The Killing House will fire around 5,000 rounds, most of which end up buried in the walls. Over time, instructors can suffer from lead poisoning. Now a new type of wall has been designed composed of angled rubber baffles that absorb the impact of the bullet, slow it down and channel it to the floor where it can be swept up after the exercise.

All SAS men are given a basic six-week course in marksmanship. This leads to a high degree of proficiency which means one man can fire a full clip of thirteen bullets from a Browning 9mm automatic pistol in under three seconds. This is four times the manufacturer's claimed rate of fire, and as each shot has to be precisely placed, it demonstrates a high level of competence. In the past, the SAS were trained in a method known as the 'double tap' or two-spaced single shots fired at a target. This technique was acceptable when terrorists only carried guns or unsophisticated explosives. Today, terrorists use remote detonation devices that require the simple push of a button to explode a bomb. Also the SAS have learned that two shots are usually not enough to stop a determined terrorist. The training now emphasizes sustained and accurate firepower; a continuous burst of fire where the impact of the bullets keeps the terrorist's hand away from his body so that there is no opportunity to press a button.

The Killing House is high on the list of tourist attractions at Hereford and visiting VIPs are generally given a demonstration of SAS firepower and rapid entry techniques when visiting the Regiment. All the chief constables in Britain and their deputies are given regular demonstrations of the SAS at work so they understand their capabilities should they wish to call on their services in an emergency. The former Northern Ireland Secretary, Merlyn Rees, has described a visit to Hereford where he visited the Killing House and saw an SAS man sitting in a room in a chair. Another SAS man burst into the room and, like the knife-thrower in a circus, stitched bullets around the man's body that when he stood up perfectly outlined his shape.

In addition, members of the Royal family and the prime minister have been irregular visitors, in part because the Regiment often provides guards for them. The prime minister has become a particular fan. A picture that has a prominent place in the Hereford mess shows Mrs Thatcher surrounded by the SAS men responsible for the assault on the Iranian embassy in 1980, codenamed Operation Nimrod.

It was the Princes Gate incident that underlined to the SAS the value of good and accurate intelligence. During the lead up to the actual assault, the bug specialists from MI5 were supposed to drill holes in the walls and insert thin microphones and tiny cameras that would give a detailed picture of who was where inside the building. But there was insufficient detail about the construction of the building and the walls turned out to be too thick for the probes to penetrate. The result was that although the SAS had a model of the construction of the building, they did not know exactly where the terrorists were.

Since Princes Gate, the SAS have built up a comprehensive computerized database that includes such essential information as the thicknesses of doors in key buildings that are potential targets, and the design of all aircraft currently in commercial service. The database is portable, so if an incident occurs at an embassy in London the SAS team can take a computer with them and access the information. By calling up, for example, the construction of the embassy, a three-dimensional image of the interior appears on the screen. Intelligence gathered on the numbers and location of people inside the building can be added to the database. Then possible methods of entry can be suggested to the computer which can plot the best method of moving through the building, tracking lines of fire and vulnerable points. If the design of the building is not known, details of the structure such as the construction of the outside walls, the number of windows and the location of bathrooms can be put into the database. The computer can then design the interior of the building setting a probability factor for accuracy which increases as more information is added.

To test their data and to test their own theories of how terrorist assaults might be made, they use a Working Attack Team to test defence targets, such as Buckingham Palace, or the cruise liner Queen Elizabeth II. They also set and run ambushes using the weapons and tactics of the IRA. This helps them develop their own responses to possible crises.

The equipment used by the SAS is evaluated by a small Operations Research Unit which is constantly on the lookout for new equipment that will give the men that extra fraction of a second that can make the difference between a successful mission and disaster. For example, since the Iranian embassy incident in 1980, the counter-terrorist units have been completely re-equipped with personal clothing, body armour and communications equipment.

The equipment used by the SAS is generally not normal army issue and the Regiment has considerable freedom to buy what it wants from its own budget.

Currently the SAS is equipped by a British company called GD Specialist Supplies with their Integrated Personal Protection Systems. This includes a Nomex fire-retardant suit, the Armourshield GPV 25 body armour vest, a National Plastics AC100/1 composite helmet, an SF 10 respirator and a new abseil harness which was purchased after one trooper in the Princes Gate assault became snagged during a rope descent from the roof of the building. For communication, the men carry the Davies Communications CT 100 unit which fits inside respirator and helmet. It has the facility to filter out high pressure sounds such as stun grenades and gunfire, while allowing normal speech to be heard clearly. Each unit can be activated either by a microphone built into the respirator or using a large 'press to talk' button worn on the chest.

Although proficient in a wide range of arms from both east and west, the Regiment still prefers the Browning 9mm pistol for close-quarter work, the Heckler and Koch MP5 machine gun for long range and the Remington 870 pump action shotgun for blowing out door hinges or other close-quarter work. The Regiment actually uses far less exotic equipment than is generally supposed, preferring instead to think their way around problems. For example, some time ago the British Atomic Energy Authority brought down to Hereford a new type of truck that had been designed to transport radioactive materials safely. The truck had been designed to withstand all kinds of attack from rifles and high explosives. It had cost at least £250,000 to develop and it was considered to be virtually impregnable. The SAS were asked to test it.

After some consideration, one of the SAS men went to the local hardware store, bought a Black and Decker drill, some rubber tubing and a Calor Gas cyclinder. He drilled a hole through a vulnerable spot in the container's side, inserted the tubing which had been attached to the cylinder and turned on the gas. He retired after lighting the blue touch paper and the new impregnable truck disappeared in a massive explosion. But such lateral thinking can cause problems. During the Falklands War one of the SAS commanders suggested that along with bombs aimed at Port Stanley airfield, three of his men be dropped from the bomb bay of a Vulcan bomber at the same time, 10,000 feet above the islands. Complete with oxygen masks and special parachutes, the men would land behind the Argentine lines to send back valuable intelligence. The suggestion caused consternation among the air staff back in London and was firmly vetoed.

The SAS do use various devices, such as night vision equipment and lasers, that are not standard issue. In particular, their communications equipment is state of the art. A special signals unit, known as 264 Signals, is attached to the SAS from the Royal Corps of Signals. They provide the satellite communication equipment that the SAS use. Recently the Regiment has been issued with a new unit made by MEL called the PRC 319, which allows a man to communicate with headquarters in Britain from anywhere in the world using a digitized high speed signal from a handset measuring twelve by six inches.

Air transport is provided by an RAF special forces squadron based at RAF Lyneham in Wiltshire. These fly specially converted Hercules transports that have systems designed to allow them to fly under enemy radar at night. Special parachutes allow the SAS to jump at heights of up to 500 feet. The Hercules also uses a technique called the Fulton Surface-to-Air Recovery System to pick up men from the ground without the aircraft landing. The aircraft drops a special kit consisting of a helium balloon with a locator beacon, two harnesses and a flexible rope that allows the aircraft to pick up either two men or a 500-pound package. On the first pass, the team on the ground pick up the kit, send up the balloon and attach the harness. On the second pass, guided by the beacon the Hercules flies in, catches the rope in a yoke at the front of the aircraft and lifts the men off the ground. By all accounts, this is a very uncomfortable way of getting home. A Hercules is kept on permanent standby at Lyneham in case the SAS have to fly off on a hostage rescue mission. The Army Air Corps also supplies some specially equipped Lynx and Puma helicopters.

There is a regular exchange programme with other counter-terrorist units in the west such as Delta Force in the United States and GSG9 in West Germany. In a crisis the SAS will frequently send observers to see another group in action and then report back on any lessons that can be learned. More often than not, it is the SAS who teach the lessons, answering regular calls for assistance and advice around the world. There is also an exchange of information on equipment and tactics. The only western power that the SAS are banned from talking to are the Israelis.

In 1987 a new command and control system was set up at the Duke of York's Barracks in London which is designed to co-ordinate all British special forces under a Brigadier with a colonel as second in command and an eighteen-strong staff. This new system is intended to cut through the bureaucracy that hampered some special forces operations in the Falklands War and also to avoid duplication of training, roles and equipment. The SAS commander in turn reports to and is sometimes instructed by the Joint Operations Centre in the Ministry of Defence. The JOC acts as a clearing house for joint operations, perhaps involving the intervention of the SAS abroad. The JOC, which is based inside the Ministry of Defence in London, includes representatives of the SAS, the intelligence services, the Foreign Office, the Home Office and various branches of the armed forces.

From there the chain of command goes through the Secretary of State for Defence to the Cabinet and, if necessary, the prime minister. The prime minister is rarely informed about SAS operations in Northern Ireland. But she was closely involved before authority was given for the SAS to assault the Iranian Embassy in London. She was also informed of and authorized the mission to arrest the three IRA terrorists in Gibraltar.

Despite the publicity the SAS has received in its counter-terrorism role, most of its training is still devoted to its wartime role – the destruction of strategic targets behind enemy lines such as communication centres, ammunition and fuel dumps and command headquarters. A Territorial SAS regiment which trains with the regulars has been given a 'guerrilla role'. It would be its job to stay behind in allied territory overrun by an enemy to carry out acts of sabotage, assassination and psychological warfare.

The SAS has become the most proficient of the world's special forces because it has gained so much practical experience in peacetime, countering terrorism around the world. It is the war against the IRA in Northern Ireland that has provided much of that experience.

CHAPTER 5

Early Days

Élitism generates factionalism and polarization within a body supposedly bonded by a common cause. The arrival of the SAS in Northern Ireland was no exception. Though steeped in the tradition of counter terrorism in far flung places, the SAS dispatched by Wilson had never been deployed against terrorists at home. Its three years of training to provide the cutting edge against international terrorism had sucked up resources and created a *corps d'élite* within the British army. The SAS's arrival in Ulster brought to the surface existing rivalries between different branches of the security forces and created recalcitrance and even enmity. Those who had fought to keep the lid on Ulster, inside both the regular army and the Royal Ulster Constabulary, either welcomed the deployment of the SAS, or begrudged the intrusion, considering their presence as condemnation of their own efforts to stem the tide of violence.

Some officers looked upon the SAS as arrogant upstarts, over trained and under disciplined – the normal regulations of a garrison parade ground cut little ice with the troopers from Hereford who knew their officers by their first names and paid scant attention to mess rules or dress codes. Equally RUC men who looked upon Ulster as their own turf were aggrieved. Those who had often lamented, 'just give us the weapons and the means and we'll sort out the IRA', had first suffered the indignity of the army invading their territory, and now had to suffer the presence of the obviously better qualified men from 22 SAS. But these were emotions, irrational responses born of frustration. Careful analysis could only conclude that the war against the IRA was now nearly eight years old, but the strategy and tactics used by the security forces remained disorganized and nowhere more than in the area of intelligence.

Since their initial deployment to the province the army had controlled policing, with the Royal Ulster Constabulary playing a supporting role. Naturally this was unpopular with the RUC, but for their part the army found much of the prejudice exhibited by the overwhelmingly Protestant RUC difficult to comprehend. Perhaps the few Glaswegians among the troops could, with their cultural background, understand the sectarian gulf in Ulster, but the hatred spawned there by religion still baffles the rest of Britain today.

There was also a widely held feeling that the RUC were not competent to carry out the task of combatting the IRA. On the covert side, the responsibility for running agents, psychological operations and undercover missions was primarily held by the Security Service or MI5. In the Irish Republic, it was the Secret Intelligence Service or MI6 who led the campaign. It has been generally accepted that there is regular feuding between the two branches of British intelligence and it is hardly surprising that they too co-operate only when necessary and only then begrudgingly. But it would be wrong to suggest that they were constantly at each other's throats. On the contrary, Ulster and the IRA threat had indeed necessitated some level of co-ordination.

However, the war in Ireland, both north and south, was perhaps the most likely to drive a wedge between the two groups. The IRA had an extensive logistics tail in the Republic. It was there they hid their stockpile of weapons and laundered much of their cash. They also regularly launched operations from the south or moved over the border after committing an act of terrorism in the north. Under such circumstances, and while there was a division of responsibilities between MI5 and MI6, there was bound to be confusion.

It wasn't just the better known security groupings that the SAS now found themselves involved with in Ulster – there was also the RUC Special Branch, which had for many years cultivated informers and agents in the community. They had the advantage of knowing people on the ground and controlling a network of narks; but the complex family and religious structures of Northern Ireland made the army wary. 'Intelligence generally was piss poor,' said one army officer who served in Northern Ireland at that time. 'We simply failed to get a grip of the whole intelligence scene to realize it was the key to fighting the war.'

As a result, in 1972, the army established its own secret intelligence gathering unit. Initially, the training was done as part of general training of troops going to Northern Ireland. This was known as NITAT for Northern Ireland training teams and involved a course several weeks long that included instruction on riot control, intelligence gathering and IRA tactics. This evolved into a more specialist covert unit. This was given the cover name of the 14th Intelligence unit or 14th Int for short. The fifty men in the unit, whose existence has remained secret until revealed to the authors during their research, was trained in part by the SAS at a secret camp near the main SAS base in Hereford. Recruits to 14th Int were taken from regular army regiments and put through a course that lasted several weeks. There was a selection process designed to test endurance and initiative similiar to that run for SAS recruits. However, while the SAS training lasts a year and continues throughout the soldier's service with the Regiment, members of 14th Int were simply taught the fairly elementary techniques of covert warfare. Recruits were taught covert surveillance, communications and agent running. The course was set up with the help of the SAS but outsiders from the security service and the Home Office also came to give lectures. The base is still used as a training centre for many of the groups that are involved in covert work outside the strict ranks of the Regiment.

The new unit preceded the SAS's arrival in Ulster. 14th Int was first sent to Northern Ireland in 1974 to operate mainly in Belfast and Londonderry. The unit used SAS methods and on occasion advisers went over to the province to see how their pupils were making out. In part this accounts for the extraordinary confusion that has surrounded the role of the SAS in Northern Ireland. For at least five years before they were officially deployed, the IRA were convinced that the Regiment was in the area and in action against them. Any covert operation that involved shooting and/or ambush was generally put down to the SAS. This has always been one of the great strengths of the Regiment: the enemy believe they are far stronger than they really are. The IRA, for example, have always maintained that when Harold Wilson first sent them in to Northern Ireland, 150 were deployed, which was nearly three times the real strength. Just why the IRA were convinced they were battling the SAS five years earlier than really were, or why they imagined so many had been deployed, has never been fully explained. It is true to say, however, that the IRA had developed a deeply paranoid fear of 22 SAS that had been exaggerated almost as soon as the troubles themselves began. It is more probable than not that a former British army NCO who joined the IRA's ranks in 1969/70 speculated or predicted that the covert expertise of the SAS might be deployed against the terrorists, and what was then no more than a rumour quickly became accepted fact. The British army has never sought to disabuse the IRA of its conviction that the SAS lies in wait around every corner and under every hillock – such fear has its place in the armoury of psyops – psychological warfare – and it has been an effective weapon.

In 1974, for example, the IRA alleged that an eighteen-year-old Pakistani tea boy working in an army canteen had been shot dead by them near Silverbridge, Co. Armagh. The IRA claimed he was a member of the SAS, which he was not. In July a twenty-one-year-old lorry driver was tortured and killed by the IRA in West Belfast. The IRA claimed he was an SAS undercover agent – which was also untrue, although he had served in the army and resigned two months before he was killed.

The terrorists were always quick to feed their own paranoia. One episode that marred the British army presence in Ulster was also laid at the door of Hereford. On 7 August 1974, Patrick McElhone was shot dead in a hayfield at Limehill in County Tyrone. McElhone, who was twenty-five years old, had no known connection with any terrorist organization. He had been taken from his home by a twelve-strong army patrol from the Royal Regiment of Wales and escorted into an adjoining field. His father followed him and had turned back to tell his wife what was happening when a shot rang out. He returned to the field to find his son dead. Lance Corporal Roy Jones was tried for the murder of McElhone and acquitted. He told the court that the unit had been under orders to arrest McElhone. When he reached the field he tried to run away and was shot dead.

The killing and the acquittal were widely condemned and the IRA insisted that it was not a member of the Royal Regiment of Wales but an SAS man secretly attached to the unit who had 'executed' the innocent man. It was paranoia or propaganda; the SAS never operates with regular units and certainly does not go out on patrol with them.

Against this backdrop, when the SAS did finally arrive on Ulster soil, 14th Int was probably the only component of the intelligence community to welcome them. Most saw it as a comment on their own poor performance, and yet more were convinced it was a political error likely to rebound on the government; for the SAS were dangerous men to let loose in a sensitive environment where the hearts and minds of the people had to be won over to the anti-republican cause. It is a measure of the army's complete misunderstanding of the role of the Regiment at that time that from their base in Bessbrook Mill they were sent out on eight-and ten-man patrols like a regular unit. 'It seemed as if the army thought that simply by being in the SAS you could do and see things on a regular patrol that would be missed by the normal squaddie,' said one former SAS officer who served in Ulster at the time. 'Our skills were operating in small units and carrying out long-term surveillance and intelligence gathering, but no one understood that or perhaps no one wanted to understand that. It was very frustrating.'

That frustration was matched by growing dissatisfaction within the security forces that they were fighting with their hands tied. The terrorists were getting away with murder, using the law and false alibis to keep them scot-free in the community under the very noses of those who wanted to hit back. The 'kill ratio' in 1975 had been running at about fifty to one in favour of the IRA. Although there had been a loose interpretation of the law by some members of the security services in the early 1970s, it was not until the SAS arrived that a more aggressive stance was taken.

The SAS were fresh from their war in Oman and had learned that the use of initiative and creative thinking helped take the war to the terrorists and put them on the defensive. It was understandable that at a time when few in the army hierarchy really understood what the SAS were about and when the rules of engagement were loosely defined, the SAS interpreted all the laws as freely as possible. The situation was heightened by a lack of proper control from above. However much the army command wanted to restrict them to conventional methods of counter-terrorism, some members of the Regiment saw their role differently and were prepared to do what they thought was necessary to put the IRA on the defensive. For men who had been used to fighting behind enemy lines in desert and jungle, the Irish border was an invisible technicality to be ignored.

This was demonstrated on the night of 11 March 1976 when twenty-three-year-old Sean McKenna returned home to his rented cottage at Edentubber in County Louth just over the border in the Irish Republic. He was sound asleep when he heard a noise in his room and woke to find two men dressed in civilian clothes standing by his bed. He was ordered to get up and walk ahead of them. Outside the house, they met up with a third man and walked across fields over the border. He was then handed over to the RUC. McKenna alleged that the men were SAS and that after they had crossed the border they put on combat jackets. He further claimed that he was beaten up by the RUC and forced to sign a statement which was used when he was later tried on a number of charges including attempted murder and bombing offences.

The SAS commander of the unit that arrested him said that his patrol were in a field on the northern side of the border looking for observation posts when McKenna stumbled across them, 'apparently drunk, rocking and rolling all over the place'.

According to men involved in covert operations at the time, the truth was rather different. McKenna was a hardened terrorist well known to the security forces. He had been aquitted in two separate trials for murder in the previous four years and he was exactly the kind of person who felt he was safe from the security forces. The SAS men did indeed go over the border and McKenna found himself believing his own propaganda, according to an SAS source involved in the operation: 'He was sure he was about to be shot when he was told to get dressed and go outside. He couldn't stop talking and gave away everything he knew without having been asked a single question. When he was handed over to the RUC he was genuinely astonished and delighted to see them.' At his subsequent trial, McKenna was found guilty on twenty-five different charges and sentenced to a total of 303 years in jail.

It baffled McKenna then, just as it must baffle shoot-to-kill theorists now, that he wasn't simply silenced in the Republic, where his death would have remained a mystery. Certainly the SAS men were not observed approaching or leaving his rented cottage, so had no reason to be 'on their best behaviour'. It would have been easier and have had a greater psychological effect to shoot him; but the SAS did not. McKenna was among the first of many to be arrested by the SAS in the coming years. A basic principle appeared to govern SAS actions; terrorists who were considered to be armed and dangerous were shot, those known to be unarmed and compliant were arrested. But there were exceptions to muddy the waters between justifiable homicide and murder.

Just over a month later, army intelligence heard that Peter Cleary, a so-called staff officer in the first battalion of the South Armagh Brigade of the IRA, was expected to visit the house of his fiancée, Shirley Hume, in Forkhill in the North. Clearly, ostensibly a scrap dealer from Beleeks, had been on the run for over a year and was wanted for questioning for a string of bombings and murders. An SAS unit kept the house under observation and on the night of Thursday 15 April 1976 Cleary was seen going into the cottage. The house was surrounded and two men with blackened faces burst in and took Cleary outside. As they left the house one of the men turned to Shirley Hume and said: 'There won't be a wedding now.'

At the inquest into Cleary's death, the SAS men wore dark glasses, navy anoraks and polo-neck sweaters and sat with their backs to the court. Soldier A testified that as they were waiting in a nearby field for a helicopter to collect him, Cleary launched himself at him. 'His arms were directed towards my throat in an unprovoked attack. I was alone with a known IRA man who earlier had attempted to escape. As he lurched at me, my instinct as an SAS soldier took over. I released the safety catch on my weapon and started shooting. There was no chance to warn Cleary. I kept on firing until the danger to me was over.'

Cleary was hit by three bullets in the chest and died instantly.

With the arrest of McKenna and the shooting of Cleary, the IRA propaganda machine was running at full steam. The Cleary killing was described as 'a cold-blooded outrage, typical of British Army operations in Ulster.' The SAS were rapidly entering local folklore as the unit responsible for every unexplained action that occurred in the province. For example, two weeks after Cleary's death Seamus Ludlow left the Lisdoo Arms in Dundalk in the Republic at around 11.30 p.m. and walked towards Smiths Garage where he tried to hitch a lift home. He was found the next morning lying by the side of the road just inside the Irish border with three gunshot wounds in his body. The IRA claimed that he had been shot by the SAS in mistake for an IRA man from Belfast who was living in a flat near his home. In fact, he was an informer for British intelligence and had been shot by the IRA themselves. But it was politically convenient to blame the killing on the SAS, which fuelled the anti-British propaganda and avoided the IRA having to admit to informers in its own ranks.

In the space of four months the SAS had succeeded in taking the initiative against the terrorists and gaining a certain notoriety both inside and outside the security forces. They had enjoyed the benefit of the doubt outside republican circles. But suddenly, one incident occurred which was to blacken their record with the army and the political establishment and raise once more fears that they were 'cowboys' operating in a sensitive area.

At 10.40 p.m. on the night of 5 May 1976 two SAS men in civilian clothes were arrested at a joint Garda/Army checkpoint on Flagstaff Road, 600 yards inside the Irish Republic. They had driven their Armagh-registered Triumph Toledo from their Bessbrook base in an operation to snatch a suspected IRA man.

The mission then went badly wrong. Normal army maps are large and too cumbersome to use on a covert operation. Instead the men were issued with a one-inch-to-one-mile map and they planned their route along the border expecting to come to a T junction marked on the map where they would turn left. In fact, the map was hopelessly out of date and the T junction had long since disappeared. Unknown to them, the men drove over the border into the south.

According to Merlyn Rees, the then Northern Ireland secretary, the Irish authorities had assured him that 'any accidental border crossing by our soldiers would not cause problems'. This agreement had clearly not reached the checkpoint, as six other SAS men in two cars sent to rescue their colleagues who tried to cross at 2.05 a.m. were also arrested.

A furious row broke out between the SAS men and the Garda, with the leader of the SAS insisting that if the roles were reversed the Garda would have been allowed to return to Irish territory without comment. 'We are all doing one bloody job,' he protested.

In the cars the police found Browning automatic pistols, two Sterling sub-machine-guns, a pump-action shotgun, ammunition, various knives and maps showing the location of farmhouses north and south of the border. The eight men were taken to Dundalk police station and then on to Dublin. Despite the intervention of the British ambassador in Dublin, who tried and failed to get an interview with Garret Fitzgerald, the then Taoiseach – Prime Minister of the Republic – the men were charged with possessing weapons and ammunition with intent to endanger life and with having weapons and firearms without certificates. The men were released on £40,000 bail and flown out by British army helicopters from the headquarters of the Irish Air Corps at Baldonnel.

Although the names of the men were given in court, they were in fact all false so that there would be no possibility of the IRA identifying them at a later date. In their trial a year later, each of the eight was found guilty of being in possession of unlicensed arms and fined £100. They were aquitted of having arms with intent to endanger life.

The operations of the Regiment became further confused that July when two men, labourer Kevin Burns and furniture dealer Patrick Mooney, were arrested at Flurry Bridge, County Louth. The two men were arrested by soldiers in plain clothes, taken to a helicopter and flown for questioning to Bessbrook. Both men claimed they had been inside the borders of the Republic and that on the flight to the army base the soldiers had threatened to throw Burns out of the helicopter.

Although the Army officially said the men involved were from the 3rd Parachute Regiment, army sources say they were in fact SAS. It is claimed the men were arrested in the north although in that part of Northern Ireland there are no border markings and the line where the border occurs is largely a matter of opinion. Certainly, the SAS team believed they were still in the north. They were in the area in the first place because they believed terrorists belonging to the late Peter Cleary's active service unit were in the vicinity. When the two men were spotted in the dark the SAS troopers believed two IRA members had walked into their trap and they were arrested. After being interrogated for three hours, both innocent men were released.

The SAS' first year in Northern Ireland had closed with some notable successes. Although they had arrived in the province to face a good deal of local hostility from other members of the security forces, they had managed to establish a clear operating pattern. Deployed in conventional large patrols when they first arrived, they soon had their normal four-man surveillance teams in operation throughout South Armagh. While their numbers remained small, the attack on Cleary and the abduction of McKenna served as clear warning to the IRA that a different war was now being fought, a war where the IRA could no longer count on the Republic as a safe haven. They were also facing an opponent as ruthless as the terrorists themselves who would fight a dirty war that, despite certain legal constraints, the SAS were better trained and equipped to fight.

But there had also been mistakes; and a body of conventional military wisdom believed the opening shots of the SAS had merely confirmed their view that prime minister Harold Wilson had unleashed a bunch of trigger-happy musclemen with no concern for legal niceties. In this, at least, they shared the opinions of the IRA. It was a false impression, created largely by the IRA's paranoia, partly by the regular forces' injured pride and, on occasions, by the Regiment's own mistakes. But it was an impression that would last, fuelled by further events – some which were undoubtedly SAS operations and others that were not. The SAS were to become victims of their own mystique and reputation.

CHAPTER 6

Mavericks

Two incidents in which the IRA and the SAS names were linked in 1977 illustrate the fog of confusion that hangs over covert special forces operations in Ulster. Both have a part to play in subsequent claims that the SAS has operated a shoot-to-kill policy in Northern Ireland.

The first came early in the New Year – a classic SAS operation involving good intelligence and covert surveillance. The army had received information that a car parked in a lane at Culderry in Armagh close to the border was to be used in an IRA operation. On 16 January, the SAS sent out a four-man patrol commanded by a sergeant which hid itself in the undergrowth nearby and settled down to await the arrival of the IRA. At around 2.15 in the afternoon the patience of the SAS men was rewarded. They heard a car draw up just out of sight. A door opened and then footsteps walked towards them. Round the corner came a man in boots, green combat clothing, with a black hood round his neck and a belt of shotgun cartridges round his waist. He also had a sawn-off shotgun in his hands.

As he moved towards the car two of the SAS men stepped from cover and moved towards Seamus Harvey to arrest him. Suddenly eight rifle shots rang out as IRA men covering Harvey opened fire. The SAS men ran for cover and in the following gun battle the SAS men fired twenty-eight shots. Harvey was killed, hit by thirteen bullets, two of them from his own men providing covering fire.

A peculiar propaganda war broke out. The army made no secret of the fact that it was the SAS who carried out the operation, while the IRA preferred, counter to its usual response of SAS brutality, inexplicably to blame a foot patrol from the Royal Highland Fusiliers. Confidence in undercover operations was growing. The security forces believed that they were at last beginning to take the initiative and that the IRA, if not defeated, might at last be fighting a more equal foe. Perhaps the IRA had decided the episode did not fit the propaganda mould it had carved out for the SAS: after all, the scene had not been littered with IRA dead whom the terrorists could claim had been mown down without mercy; and all the evidence pointed to the SAS attempting to arrest their armed suspect until the hidden active service unit opened fire.

But then twenty-nine-year-old Grenadier Guardsman, Captain Robert Nairac, was abducted, tortured and killed by the IRA. To this day it is suggested that Nairac was an SAS man operating behind the lines and quietly assassinating leading IRA terrorists; his name is regularly cited in the shoot-to-kill debate.

Nairac was the ideal army officer. He joined the Grenadier Guards after being educated at Ampleforth, the top Catholic public school, and Lincoln College, Oxford, where he read history. He was both intelligent – three A levels and nine O levels – and a good sportsman with a boxing blue, a marksman's certificate and a reputation as a keen and proficient falconer. After Sandhurst, he had served a number of tours in Northern Ireland until in 1977 he was based at Portadown, spending several nights each week at Bessbrook Mill in County Armagh. It was widely rumoured at the time, and has been ever since, that he was a member of the SAS. In fact he was never a member of the SAS; he was seconded to the army's under-cover intelligence gathering unit, 14th Int.

In that capacity, Nairac had developed a Belfast accent which he believed he could speak fluently and enjoyed going to local pubs in his Triumph Dolomite to chat to the locals and join in the evening singalongs. He was well known in the area and called affectionately Danny or Danny Boy because of his fondness for that song.

It is a measure of the naïvety of the army at this time that they thought that an army officer could simply walk out into the streets and be absorbed in the community, particularly an officer as classically British as Nairac. Some sources in the IRA claim that his Irish accent was so bad that he had become well known as a British officer, but was tolerated because he was so obvious that he was thought to be a plant. Alternatively, it was felt he might be a 'Stickie', as members of the Official IRA are known.

Nairac rightly believed that the key to British success against the IRA was good intelligence and he also believed that he could provide it on his forays out into the community. To those who knew him, Nairac was an enthusiastic and intelligent officer with an overly romantic view of combating terrorism. He enjoyed the idea of mingling with the locals and playing the role of a dashing undercover warrior but did not have the necessary cynicism, modesty and thoroughness so essential to sur-vival in such a job.

Standard operating procedure for any covert operation says that an undercover operator should never go out without adequate back-up and should certainly never be out of touch with base. But on the night of 14 May, Nairac went to the Three Steps Inn at Drumintee in County Armagh. The pub is isolated, on a lonely hill out of range of any immediate help and three miles from the border with the Republic. In any event, Nairac, dressed in jeans and anorak, had arranged only to call in to his base at 11.30 that evening. In a real emergency he could press a distress button hidden inside his car which could have summoned help if any was within range. In addition, he was carrying a Browning 9mm pistol concealed in a shoulder holster.

The pub was packed that night with around 200 people, and after an evening of steady drinking the songs began. Nairac had been on stage and finished his version of two favourite IRA tunes, 'The Broad Black Brimmer' and 'The Boys of the Old Brigade'. He left the pub and got into his car to drive back to base when it was surrounded by nine members of a local IRA terrorist unit.

Before he had time to draw his pistol, the men had begun to drag him from the car. In the fight that followed the car windscreen, wing mirror and aerial were all broken and the car dented in a number of places. Nairac was eventually knocked unconscious, bundled into a car and immediately taken south over the nearby border. For a short time he was left unconscious in a house with only a single guard. Nairac recovered and attacked his guard, knocking him down. He grabbed the guard's revolver and as another IRA man rushed into the room pointed the pistol at him and pulled the trigger. The weapon misfired. He pulled the trigger again and it misfired once more, he pulled the trigger a third time and once again it misfired. By then, the original guard he had attacked moved behind him and knocked him unconscious.

'It was a tragedy', said one former army officer who served with Nairac. 'He was so near to getting away and then to have a weapon misfire three times. The chances of having three stoppages in succession are unbelievable.'

Nairac was swiftly moved from the house and taken into Ravensdale Forest near the border. Beside a bridge he was bundled out of the car and for several hours brutally tortured by his captors who wanted to find out details of

SAS operations in the area. Knowing he was about to die, Nairac, a Catholic, asked his captors for a priest. In a final and macabre humiliation, one of the terrorists played the part of the priest to hear the officer's confession. He was then killed with one shot to the head and another to his body; his corpse was never recovered, and according to Republican rumour was fed into an animal-feed plant.

Nairac had failed to make his 11.30 p.m. call back to base at Bessbrook but the alarm was not raised until 6 a.m., by which time he was very probably dead. Although it is very unlikely that Nairac could have been rescued, the length of time it took the army to respond to what was clearly an emergency showed a lack of appreciation of the situation. Despite an extensive search by the security forces no trace of Nairac's body was found. Two days later, the IRA issued a statement saying that he had been killed. 'We arrested him on Saturday night and executed him after interrogation in which he admitted he was in an SAS unit. Our intelligence department had a number of photographs in their possession and the late Captain had been recognized from them.' In fact, despite being tortured in the most brutal manner, Nairac never told the terrorists anything.

Ironically, it was the IRA's fear of the SAS that was to lead police to Nairac's killers. Later in 1977 a twenty-four-year-old joiner from Meigh in County Armagh, Liam Patrick Townson, was arrested by the Irish police on suspicion of being involved in Nairac's murder. He was closely questioned about the shooting and according to the Garda drew a detailed sketch of the scene of the killing. He pointed out where the terrorists had hidden two guns – including Nairac's Browning – as well as the officer's clothing in the forest.

The Garda also claim that Townson told them: 'I shot the British captain. He never told us anything – he was a great soldier.'

At his subsequent trial, Townson denied making the confession and claimed he had only taken the police to the caches in the forest when the Gardai bundled him in a car and began to drive towards the border. 'I said to myself: if I don't show them where these bags are, I'm going to end up a dead man. I said I would probably end up at the back of a ditch with a hole in my head if I was handed over to the SAS.'

Townson was jailed for life for his part in Nairac's killing, and information he supplied to police led to the arrest of five other men in Northern Ireland who were jailed the following year for their part in the murder. In February 1978, two months after the five IRA terrorists had been jailed, Bob Nairac was awarded the George Cross, the highest peacetime honour a serviceman can receive. The citation spoke of his exceptional courage and 'acts of the greatest heroism. In circumstances of extreme peril, he showed devotion to duty and personal courage second to none.' The citation added that Nairac had been subjected to particularly brutal assaults in an attempt to extract information but he had refused to disclose any operational details to his captors and had made repeated attempts to escape.

Nairac was a popular hero for a war that badly needed them. There was no doubting his courage or his devotion to duty. But his death caused the army to conduct a re-examination of the way covert operations were carried out. It was decided that there would be no more freelance operations; that no operative would go out without sufficient support. New guidelines were drawn up to improve the response time of back-up forces if undercover agents failed to make their check calls on time.

But Nairac's name surfaced once more in 1984, to fuel the undiminished suspicion that British forces and the SAS in particular were deliberately shooting to kill, even assassinating, terrorists. Seven years after Nairac was killed allegations linked him to the 1975 murder of John Francis Green, an IRA commander in North Armagh who was shot dead at a remote mountainside farmhouse in County Monaghan in the Republic. Nairac is alleged to have boasted of killing Green and to have had a photograph of his dead body. Nairac is also alleged to have been involved and had boasted how he went over the border in the shooting of members of the Miami Showband in 1975. The RUC has always maintained that the UVF, the Protestant paramilitary group, was responsible for the Miami Showband killings – something the UVF have themselves admitted. Three members of the UDR who were also members of the UVF have been jailed for the murders. The police suspect that the same extremists were responsible for Green's murder.

The allegations found a champion in the House of Commons. Ken Livingstone, in his maiden speech as an MP, repeated the charges that Nairac had been behind the Miami Showband killings and the murder of Green. The allegations remain unsubstantiated and the RUC remains convinced by the weight of available evidence that the Protestant paramilitaries were behind both killings. It is true that Nairac had a picture of Green's dead body, but this was obtained from a contact in the RUC who in turn obtained it from the Irish police.

The allegations against Nairac illustrate some of the wilder comments that have been made over the years about the British security forces' policy in the province. And to this day there are many who believe, despite the overwhelming lack of evidence, that Nairac was an SAS maverick actively assassinating terrorist suspects. It is an enduring element of the shoot-to-kill legend. This has never been a policy dictated by politicians or army commanders; no one needs convincing that such an action would be totally counter-productive, as the controversy itself has shown.

It is a fundamental rule of counter-terrorist theory that to combat terrorism effectively, government forces must adhere to the rule of law. If it reduces the war to a no-holds-barred slugging match, the government loses the moral high ground and plays into the hands of the terrorist propaganda machine. The top IRA men all know that they have been identified and are easily accessible to the security forces should they wish to kill them. Their freedom is testimony to the nonsense of a shoot-to-kill policy. The war against the IRA is waged according to unwritten rules. The most basic is that as long as the security forces remain within the law, terrorism cannot win. Government resolve will outlast a guerrilla army's cause, as long as it is seen to be upholding the law. If it once abandons the constraint of law, government risks losing the support of the population and transferring it to the terrorists themselves. It is precisely for this reason that the IRA propaganda machine goes to such great lengths to fuel the shoot-to-kill controversy.

There is another unwritten rule unlikely ever to be debated. Those opposed to the IRA, including soldiers, policemen and politicians, regard themselves as legitimate terrorist targets, and while the IRA restrains itself to atrocities against such people, including the civilians who might support the forces of law, it remains both subject to and protected by those very laws. It is hard to imagine an act or strategy that the IRA could commit itself to which might change that, but a shoot-to-kill policy is a 'final solution', an ultimate sanction that the government holds over the IRA leaders it could so easily identify and remove. As one former army officer with extensive experience in Northern Ireland said: 'If an unwritten rule is put into play such as: stay with legitimate targets and we will play the game but go outside those then it is open season on the key men. This can have a useful effect of serving to contain the mad dogs.'

However, while murder is not tolerated, in the 1970s

there was often fairly free interpretation of the rules of arrest. Also, because the Republic's Garda were frequently unhelpful in acting on information supplied by the security forces and even on occasion actively helped terrorists on the run, crossing the border was considered a legitimate tactic. As relations between north and south improved so the freedom of action of the SAS and other undercover units became more constrained.

But while the army may have recognized the need to operate inside the law, some members of the RUC felt less inhibited. The RUC had never liked the British intelligence operating on what they considered their patch, and the arrival of the SAS exacerbated their chauvinist fears. If the SAS and 14th Int could get involved in agent running, intelligence gathering and armed operations then, they argued, so should the RUC. They proposed that a new unit, known as E4A, be established which would act as the covert intelligence-gathering arm of the RUC working to complement the Special Branch. The Special Support Unit, SSU, also known as the HQ Mobile Support Unit, was established as the armed 'active measures' unit.

The Deputy Chief Constable of the RUC, Michael McAtamney, giving evidence at a later trial of one of the SSU men, said they were trained in 'firepower, speed and aggression'. He added that while there were clear rules of engagement an officer certainly would be governed by the qualification that minimum force should be used but, at the same time, the objective of his training is to 'eliminate the threat.'

The men of SSU, who numbered around thirty, were trained by the SAS at their base in Hereford. But the SAS did not select the recruits and there was unease in the Regiment at the calibre of men they were required to train. Undercover work requires a dispassionate commitment, where training, intelligence and initiative combine to produce a man who makes calm and rational decisions under stress. To the SAS the RUC men appeared overly committed, gossipy and passionate in their fight against the terrorists.

The SAS' reservations about the unit, expressed at the time but ignored, were to prove justified.

On 27 October 1982, in a massive landmine explosion, the IRA blew up a car carrying three members of the RUC in Lurgan. It was a brutal attack in which all three policemen were killed. For the RUC the explosion proved the final blow. They had seen their colleagues bombed and shot and their response was dictated by the letter of the law. Many of the terrorists operating in the province were known to the police but arresting them was impossible as there was insufficient evidence that would stand up in court. Of course the IRA recognized this perfectly well and were happy to hide behind the law at every opportunity.

Although no formal shoot-to-kill policy was ever adopted in the sense of instructions coming down from the Chief Constable ordering the elimination without trial of IRA suspects, there is no doubt that some members of the RUC now adopted a less formal approach to policing than had previously existed in the province.

On 11 November 1982, Eugene Toman, Sean Burns and Gervaise McKerr were shot dead in their car on Tullygally East Road just outside Lurgan. The RUC claimed that the men were shot at a police road block after their car had failed to stop, and in driving through the roadblock had knocked down a policeman. In fact, the three men had been under surveillance for some time as it was suspected they were involved with the IRA. They drove into an ambush and the police fired 108 bullets into the car from rifles, a Sterling sub-machine gun and a pistol.

On 24 November two youths, seventeen-year-old

Michael Tighe and nineteen-year-old Martin McCauley were shot by police as they climbed into a hayshed outside Lurgan. Although the two youths had no known connection with a terrorist organization they were going into a barn that was under surveillance both by the SSU and through a bug that had been planted by British intelligence.

Then on 12 December 1982 Seamus Grew and Roddy Carroll were shot dead by men from the SSU. Neither of them was armed.

Relations between the British government and the RUC have always been fragile, and although the killings had been noticed there was little enthusiasm in London for trying to stop them. Through their own sources, the SAS had heard that the men they had trained had taken the law into their own hands.

Even within the RUC the more vigorous approach to policing was recognized by more senior officers as an exceptionally dangerous course. They understood that if it was allowed to continue then the rule of law would break down totally in the province; and in fighting terrorism it was only the rule of law and its strict application that had distinguished the security forces from the terrorists in the eyes of many in the Catholic community.

Two years after the shootings, when a number of policemen had been tried and acquitted, the Deputy Chief Constable of Manchester, John Stalker, was called in to hold an independent inquiry. Over the next two years Stalker says his investigation was blocked at every turn by senior officers in the RUC. In part this was because they feared what he might find, in part because they resented an outsider poking into their affairs, and in part because it was felt his investigations and what they might reveal would be bad for the morale of the force, and in part because they believed Stalker was indiscreet.

Almost immediately after Stalker and his small team arrived in Belfast, they became victims of what they saw as a blatant attempt to cover up the killings. According to Stalker, statements made by a number of policemen were untrue and others lied to him on the instructions of their senior officers. But over the next year and a half, Stalker believed he uncovered the basis of a clear 'shoot-to-kill' policy being carried out by members of the RUC. It was not alleged that this had been ordered by the government or by the RUC Chief Constable. On the contrary, what seems to have happened is that a small number of RUC men decided to take the law into their own hands and, once their actions were discovered, the RUC decided to look after its own and thwart any outside investigation.

Just as Stalker was about to finish his investigations, allegations surfaced in Manchester that he had been associating with people involved with crime. This was a very serious charge and he was suspended from duty while the matter was investigated. The charges were found to be totally without foundation and the suspicion remains that some members of the RUC caused the allegations to be made in order to muddy the waters around the Stalker investigation.

Although the report was completed and action was recommended against a number of officers, no criminal prosecution resulted because the Director of Public Prosecutions argued that to do so would not be in the national interest. In fact, it was feared that a long and complicated trial would inevitably bring into question the leadership of the RUC, the training of its men, the extent of any conspiracy (if any) and, above all, that it would look in detail at the operating methods and intelligence of the RUC. Such a trial would provide invaluable propaganda and intelligence for the IRA and it was considered that the price was too high.

Instead, a number of policemen were quietly offered early retirement and others were suspended from duty. Perhaps most importantly the SSU, the SAS-trained team that had carried out the killings, was disbanded. Although E4A still carries out intelligence-gathering operations, the RUC no longer performs the kind of operations, such as ambushes, that might lead to shooting. That is once again left to the SAS.

It was the RUC which had argued that it should be controlling the undercover war against the IRA and it was the British government and, in particular, the British army which had agreed. In the wake of their clear failure to meet their responsibilities, the army developed a new undercover unit to carry out surveillance and intelligence gathering while at the same time strengthening the command and control structure along the border with the south.

Like E4A, the Intelligence and Security Group – ISG – was trained by the SAS at the Hereford base. Like 14th Int, their recruits come from the regular army on secondment. They are trained in covert surveillance and intelligence gathering over a period of weeks and are then attached to each of the three army groups based in Northern Ireland.

While the RUC and other undercover branches of the security forces had been learning new tactics to confront the IRA, the SAS had watched their specialist role diluted by other off-shoots of the security forces and the odd maverick operation; they had seen their reputation for ruthless efficiency both heightened and tarnished as every new questionable incident often perpetrated by others was attributed to them. But the SAS on occasions also made mistakes and created its own bad publicity.

The successful deployment of the SAS relies on three key factors: accurate and timely intelligence, good liaison with other units working in the area, and the training of the SAS men whose fast reactions are often the key to an operation. Given the knife edge on which the SAS operates it is hardly surprising that there have been mistakes. But it is prudent to point out that the errors have cost the lives of SAS men as well as terrorists and civilians.

Three incidents in particular demonstrate the lessons learnt by the SAS.

In March 1978 the British army received intelligence that terrorists were using a house at the bottom of the Glenshane Pass outside the village of Maghera in Londonderry. On the night of 16 March two SAS soldiers, Lance-Corporal David Jones, aged twenty-three, and another man were hidden in a hedgerow overlooking the house. Through their night vision glasses they saw two men coming towards them along the hedgerow and heading for the house. Both men looked as if they belonged to the UDR: they were wearing camouflage jackets with the word 'Ireland' written on their shoulders and both were carrying rifles. As they drew opposite the two SAS men, Jones stood up and, expecting a friendly reply, shouted a challenge. Immediately one of the men, whose rifle happened to be pointing at the hedgerow, opened fire. Jones was hit in the chest and the other man wounded in the stomach. Despite his wound he managed to return fire and one of the terrorists was hit while the other made good his escape. The second SAS soldier called for help on his radio and Jones was evacuated to the Mid-Ulster Hospital at Magherafelt, where he died the next day.

At dawn that day (St Patrick's Day), dogs were brought in to the area and the hunt for the terrorists resumed. One Alsatian picked up a trail of blood and his handler followed it to a nearby waterhole. 'At the waterhole I saw a black leather belt lying on the ground. The dog then went on another course from the hole and I saw blood on barbed wire in another field. The dog went on towards a house at the top of this field. Then it lost the track. However, after information from the police, I took the dog to a house and there were blood spots there.'

A few minutes later the dog led the handler to a nearby bunch of gorse bushes and he caught sight of someone lying inside them. I could see him. His hair was blondish. I said come out and the man replied: 'I can't move. My legs.'

The man was dragged from the bushes and was found to have a bullet wound in his thigh. He later needed three operations to save his leg which was shortened by an inch and a half in the process. As he was carried into a waiting ambulance he shouted: 'Up the Provos,' the last words the police got out of him for some time. Although his hair was dyed blond, police identified the man from his fingerprints as Francis Hughes, a twenty-two-year-old joiner from the village of Bellaghy in Londonderry. Hughes had been on the run for five years, mainly in the company of Dominic 'Mad Dog' McGlinchey, a prominent terrorist with a long list of kills to his name. In 1972 Hughes had set up his own terrorist unit in the South Derry area and after a string of attacks, including the murder of two policemen, he was accepted into the IRA. He operated in full combat gear and frequently rang the British army to tell them where he was, acquiring such notoriety that most IRA attacks in the area were blamed on him. Hughes died on hunger strike in 1981 telling medical staff: 'I'm dying, I'm dead. That's it.'

After every incident, particularly one involving the death of one of their own men, the SAS hold a detailed investigation to see what lessons can be learned for the future. The fact that an armed terrorist was able to open fire first on an SAS team engaged in covert surveillance clearly demonstrated that there had been some failings.

It is standard procedure for security reasons that the SAS do not inform regular army units if they are mounting a covert operation. However, all regular units are told if certain areas are out of bounds. But the SAS should be aware at all times of units from the army and police that are likely to impinge on surveillance missions. This is especially so at night when mistakes are easily made and identification is often very difficult. But in the case of the Hughes shooting, the SAS were unclear about the patrolling patterns in their area because of poor liaison between the police, the army (including the UDR) and the SAS. Steps were taken to improve this relationship which should, in theory at least, enable the SAS to differentiate more clearly between friend and foe in future.

But the death of Lance-Corporal Jones had a more lasting psychological impact on the Regiment. It was recognized (and would be repeatedly emphasized in training in the years to come) that taking chances with IRA terrorists was a deadly business. The speed of reaction of Francis Hughes had been faster than that of the SAS men, something that clearly showed the men were insufficiently prepared for the confrontation. Ten years later, when the SAS shot three IRA terrorists in Gibraltar, the question of split-second reactions would play a large part in the Regiment's explanation, for the terrorists turned out to be unarmed. The SAS shot them even before a full challenge was shouted because they had seen what they thought were threatening movements. At the time, the Regiment privately justified their fast and deadly response by pointing out what happened when they took a chance with Francis Hughes. They had learned their lesson then, they said. Why should any SAS man risk his life to give a terrorist, clearly caught in the act of aggression, the benefit of the doubt?

The first opportunity after the Jones killing to test the improved liaison system with the other members of the security forces and their responses in a crisis ended in disaster. On the afternoon of Monday 11 July 1978, sixteen-year-old John Boyle was making hay with his brother on the family farm near Dunloy in County Antrim. At around 4.00 p.m. he broke off from his work and strolled across a narrow country lane into a nearby graveyard to look for some of the family headstones. Underneath a headstone that had fallen down and was lying at an angle he saw the edge of a plastic fertilizer bag.

Inside the bag was an Armalite rifle, a revolver, an incendiary bomb, a face mask, combat jacket and black beret – all the equipment for an IRA man of war.

Excited by his find, John ran back to fetch his older brother, Hugh, who in turn went to fetch their father Con Boyle. He immediately telephoned the local police station to alert them to the terrorist cache. The police passed the information on to an army captain who was acting as liaison officer with the RUC from his base at Ebrington Barracks, Londonderry. It was decided that a four-man team from the SAS would move into the graveyard, under cover of darkness, to keep the arms under surveillance. As is usual in these cases, the army captain was briefed by the RUC and he in turn passed all relevant information to the SAS. It was here that two fatal errors were made.

First, the RUC said that the arms had been found by a boy or young man. By the time that information reached the SAS patrol, the description had been translated to 'found by a child'. Second, the RUC promised that they would speak to the family and warn them not to approach the graveyard. The military assumed that had been done immediately. In fact the family had not been contacted after Con Boyle made the initial telephone call and had pinpointed the cache. The policeman responsible for speaking to the family ended his shift at 2.30 a.m. and did not return to work until 9.30 a.m. It was only then that an attempt was made to contact the Boyles and the message did not reach Con Boyle until around 10.15 a.m. when he was working in a field about half a mile from the graveyard. By then it was too late.

Young John Boyle had already left to go and see what had happened to his graveyard hoard. In their final briefing before they left for the stake out, the SAS men were told that a wanted IRA terrorist, Eugene O'Neill, was in the area and it was suspected the arms might have been left for him. The four SAS men had taken up their positions five hours earlier. Two of them hid in a ruined farmhouse over-looking the entrance to the graveyard. The other two, Sergeant Alan Bohan, aged twenty-eight, and Corporal Ronald Temperley, twenty-six, hid in a camouflaged hide between a hedge and the stone wall enclosing the graveyard about twelve yards from the gravestone. They sat back to back with one facing the approach to the stone and the other looking directly at it.

After several hours' waiting, Bohan heard the creak of the cemetery gate. He nudged Temperley and both men watched John walking up the path towards them. He passed within three yards of the SAS hide and both men saw 'someone too old to be the child' and flicked off the safety catches of their Armalite rifles. John reached under the gravestone and, according to the SAS men, pulled out a rifle, turned in their direction and pointed the weapon. Without hesitation both men fired and John was hit by three bullets, one in the face and two in the back. He died instantly.

It was a tragic error that could have been avoided. Both the SAS men were subsequently charged with the murder of Boyle. It was alleged at their trial that they had fired on the boy without warning from behind when they could easily have arrested him. A policeman also testified that he had told the army three times that members of the Boyle family could return to the graveyard the following day. Testifying at his trial, Sergeant Bohan said he thought that John Boyle was 'someone whose picture I had seen previously'.

'When he turned to his left he was facing us full on and the rifle was raised. I was certain he was about to fire. I opened fire and his head dropped to the right as his body fell to the left so that his back was facing our operations post. I fired once or twice, I don't honestly remember. Temperley also opened fire.

'I thought I was firing on an armed terrorist who was bringing a weapon to bear on us. I gave no warning whatsoever, I thought it was impractical.'

The soldiers' evidence was contradicted by the path-ologist who argued that all three shots had entered the boy's body from behind. However, the judge accepted the SAS' evidence and aquitted both men of murder. But he described the operation as a 'badly planned and bungled exercise'. The judge said that the soldiers must have seen John pick up the rifle from the cache 'but strangely did nothing while this was going on'. If they had inspected the rifle beforehand they would have found out that it posed no threat in its unloaded state (though this would have been impractical, too, for the SAS could not risk breaking cover to inspect the cache for fear their presence might be detected).

'Bohan may have thought that the deceased was a terror-ist because he had reason to think that the boy's family had been warned about the arms cache. If he thought this, one wonders why he allowed him to take a rifle into his hands,' pondered the judge.

The Boyle killing played straight into the hands of the IRA propaganda machine and even ten years later it is regularly referred to as an example of the 'brutal shoot-to-kill' policy allegedly adopted by the SAS. It is a naïve assumption. The SAS' masters have learned how such mistakes can profit the IRA and the troopers at the sharp end are the last people in the world who would wish to give comfort to the enemy.

The Boyle shooting had not been ordered from someone on high or even at low level; it was a tragedy for the family and a lesser ordeal for the SAS. The Regiment had been deployed correctly, had reacted to circumstances which suggested to them a terrorist was in their sights and had opened fire. The Regiment itself would never have con-doned an action that might suggest they were, after all, murderous cowboys with little regard for innocence. Such incidents can be minimized by training and experience but others cannot.

The SAS' expertise is the well-planned assault or ambush which is always backed by accurate intelligence and surveillance and intelligence gathering over long periods in a hostile environment. All their training is geared to split-second responses against a clearly defined target. Such expertise is instinctive, a second nature that the SAS find difficult to suppress in the heat of battle where human strengths and weaknesses are so exposed. Their expertise is out of place in normal circumstances when regular troops would be just as effective. But on the other hand because the SAS are so highly trained, they can too easily be considered flawless supermen who can walk blindfolded into battle and walk away unscathed. That is exactly what happened on the afternoon of Friday 2 May 1980.

At around 2.00 that afternoon the army received a tip that there was suspicious activity at a house at 369 Antrim Road, Belfast. The information suggested that there might be an arms cache in the building. The normal routine would have been either to send a regular army patrol to the area or to keep the house under discreet surveillance to see who came to collect the weapons. In fact, army headquarters decided that an eight-man SAS unit in the area should go to the property.

The men were in two unmarked Morris Marina cars. Dressed in jeans and anoraks, they were armed with Heckler and Koch MP5 machine guns, Colt Commando automatic rifles and Browning automatic pistols. Each man wore a green fluorescent armband for easy identification. The message from army headquarters was received by Captain Richard Westmacott, who was sitting in the back of the lead car. The message to Westmacott told him to move with all speed to the Antrim Road address and break into the house. It was expected that arms would be there but there was no suggestion that the house was occupied at the time.

One car with four men inside went to the rear of the

house while Westmacott's car pulled up at the front. One of the soldiers inside the car now takes up the story:

'Our car stopped outside the doorway of no. 369. We all started to get out and we came under fire from a top-floor window. I moved for cover to the rear of the vehicle. I then moved across towards the doorway. We believed we were coming under fire from no. 369. I saw muzzle flashes and then breaking glass form the window to the left of the arched doorway leading to no. 369. At that stage, I would say there was more than one weapon being fired at us. Captain Westmacott was with me moving towards the door of no. 369. I reached the doorway along with two other soldiers. I was the third man in and I stayed on the ground floor to give cover while the other two went upstairs.'

Using the carefully rehearsed techniques learned at the Killing House at Hereford, the two men cleared each room in the house, one man forcing open each door while the second entered. They found no one inside; the gunmen were not in no. 369 but next door in no. 371. When the SAS men had finished clearing the house they realized that their commander was missing.

'On the pavement I saw Capt Westmacott lying there about six feet from the door. His body lay to the right of the parked car on the forecourt. I saw two people with him – one looked as if he was a doctor. I went up to the body and decided he was dead.'

In fact Westmacott had been cut down by two bullets, one in the shoulder and one in the head, as soon as he left the cover of the car. He was killed instantly.

As the shooting began the alarm had been raised and the area swiftly cordoned off by the RUC and regular army units. While the SAS returned to their base, an RUC police superintendant called on the men in the house to surrender. They replied that they would not come out until the SAS had left and warned, untruthfully, that they had a woman hostage. They pointed an M60 machine gun through the window and threatened to open fire if any attempt was made to storm the house.

After the arrival of a local priest and a further thirty minutes of discussion the four men in the house surrendered. As they were being led away, one shouted 'Up the Provos', while another bellowed: 'See you in thirty years, lads'. The four men were all known terrorists and were on the run after escaping from Belfast's Crumlin Road jail.

No unit – either SAS or regular army – should have been sent to the house in Antrim Road on the basis of hopelessly inadequate intelligence. The folly was compounded by ordering the SAS to enter the property with instructions to clear it. The result was the unnecessary death of Captain Westmacott, a highly trained soldier, lost in a firefight he should never have been asked to attend, and which could have been avoided by more sensible means.

The SAS have been involved in dozens of incidents in Northern Ireland and have lost far more men than Lance Corporal Jones and Captain Westmacott. But those two deaths and the shooting of John Boyle were important milestones in the SAS' transition to an effective anti-terrorist tool. The Regiment itself had a steep learning curve in the province. Northern Ireland was very different from Oman or Malaysia, where operations could go unscrutinized, and where rules of engagement could often be interpreted to suit the circumstances. They have had to work very closely not only with regular army units but also with the civilian police. For a body of men used to working with each other and to their own standards, such a compromise has been difficult. But they learned the hard way that liaison is vital and some measure of co-operation essential if they are not to lose more of their own men.

It has also been difficult for the regular army and the RUC to understand that the SAS are more than simple soldiers. Their very exacting training turns them into precision instruments designed for specific jobs. They are not blunt objects to be thrown at the nearest door in the hope of battering it into submission. Perhaps most important of all, in Northern Ireland the full glare of the media is permanently turned on the Regiment and any action is examined not only by the press but by an ever-efficient IRA propaganda machine.

These lessons were carried into the eighties. By then the knife edge between heroism and villainy on which the SAS operated had grown razor sharp, honed by experience, error and the increasing sophistication of the terrorist propaganda network. The more acute that edge became, the more likely it was that the SAS and the government would find not just the odd trooper under cross-examination, but the entire Regiment in the dock.

PART TWO

CHAPTER 7

Loughgall

They were only toddlers and infants when the troubles flared in 1969, two of them still in nappies, but they had grown up with violence. As children they hurled abuse, bottles and bricks at troops, in adolescence they carried messages and kept watch on street corners for the gunmen; it was only natural that, in manhood, they would progress to the ranks of the IRA to be versed in the techniques of terror. There were eight in all, four from the same village, and they comprised the sharp end of the terrorist network in this stretch of bandit country. Two active service units from the East Tyrone Brigade had joined forces for the attack, the latest and, they hoped, the most spectacular assault on the RUC. The godfathers had decreed that a sustained campaign of beatings, bombings and bullets against police stations and any builders, traders, caterers or delivery men serving the security forces' outposts would be the latest turn of the tactical screw that might one day make Ulster ungovernable.

The plan was both audacious and ambitious, hence the rare coupling of two frontline units. It had been months in preparation; the target had been chosen well in advance, the weapons and explosives had been collected from the secret caches south of the border, the vehicles to be hijacked had been selected and the bomb's hiding place prior to its use had been chosen; the go-ahead was given by Northern Command which in turn had informed the seven top brass on the Army Council. All that remained was to choose the right time for the assembled team to strike. No one realized how suicidal was the action they now planned.

James Lynagh was their leader. At thirty-two he had risen through the ranks of the IRA to command the Armagh and Monaghan border region from the relative safety of his home in the south. From Monaghan city he controlled the gunmen who carried out the great majority of assassinations north of the border in Tyrone and Armagh bandit country. His thirteen years in the IRA had earned him a ruthless and legendary reputation; such a man was a natural successor to Seamus McIlwaine, the terrorist overlord of border operations who was shot dead by the SAS in 1986. Lynagh, a hardliner, was not easily controlled by the Northern Command which found that giving him his head in the rolling countryside around the border was preferable to trying to subject him to military discipline. He and his two brothers were considered hardline mavericks; one committed suicide while serving a prison sentence in Dublin and the other, sent out to perform the kneecapping of a nightclub bouncer who had angered the IRA by beating up Jim Lynagh himself, imposed a stiffer sentence. He murdered the offender and earned a life sentence for himself.

Lynagh had been a Sinn Fein councillor in the south but in the north he figured high on the RUC's most wanted list. That he was tough, there was no doubt; he had survived a premature bomb blast in the seventies when the device he was carrying exploded in his lap. He had been jailed for ten years for possessing explosives after his recovery, which the judge described as miraculous, and, on his release, quickly resumed his murderous activities. Luck was to play a large part in Lynagh's survival.

In 1980 he was one of the first terrorists to be arrested and charged in the south for a murder in the north, under new Anglo-Irish laws. He and two others stood accused of murdering a former member of the UDR at his farm but the charges were dropped in Dublin's Special Criminal Court when three judges upheld a defence submission that the prosecution had not established a prima-facie case. The old and infirm were considered legitimate targets by Lynagh as long as service to the crown could be counted against them. In 1981 he shot dead the former Speaker of the Stormont parliament, Sir Norman Stronge who was eighty-six; his son James was also assassinated in the raid on their country home. It was no secret in the republican movement or in RUC headquarters that Lynagh was the man who had pulled the trigger but there were no witnesses to identify him and no weapons for ballistics to examine; by the time the bodies were found, Lynagh was back in the south preparing for a council meeting.

Lynagh's second in command for the raid they now planned was Patrick Kelly, thirty, who led the East Tyrone brigade of the IRA and lived in Dungannon. He was the only married man selected for the attack and his wife was expecting their fourth child. But such considerations carry little weight and Kelly was not the type to miss the opportunity to enhance his own reputation.

The third most senior terrorist recruited for the attack was Pat McKearney, thirty-two, from Armagh. He had been on the run since 1983 when he and thirty-seven other IRA men had broken out of the Maze prison where he was serving a fourteen-year sentence for possessing a loaded sub-machine gun. A fourth was Gerard O'Callaghan, twenty-nine, of Benburb. An old friend and comrade in arms of Kelly's, he had murdered a UDR soldier and an RUC officer. The four men represented the old hands on the team; there were two others who were gaining in experience, Micky Gormley and Eugene Kelly, both twenty-five, and two virtual beginners; Declan Arthurs and Seamus Donnelly both twenty-one and only a little older than the troubles themselves. The two active service units combined the frontline strength of the East Tyrone brigade under the leadership of Lynagh. It was a formidable line up of IRA hardmen.

Their target was 'soft', a part-time, undermanned RUC station in the largely peaceful village of Loughgall, a Protestant enclave in the middle of Catholic County Armagh, thirty miles south west of Belfast and a few miles from the border. The troubles had rarely touched the remote community in the heart of Ulster's beautiful orchard country, an apple-growing region populated by fruit farmers from England's Vale of Evesham centuries ago. But, surrounded by the towns of Dungannon, Armagh and Portadown, trouble had never been far from Loughgall. It had come close a year earlier when another part-time station at The Birches, five miles away, was unsuccessfully attacked by the IRA.

The attack on Loughgall was to be a carbon copy; the active service unit would steal a JCB mechanical digger which would ram the gates carrying a huge bomb in its bucket. Terrorists would ride on the digger armed with shotguns to eliminate any policemen who might get in the way and others would follow behind in a stolen van, which would both reconnoitre the area before the attack and drive them away from the scene after it. The decision to repeat the strategy was the IRA's first blunder.

As soon as the digger went missing there could be only two conclusions; it was a simple theft in which the vehicle was to be smuggled south for sale or the terrorists were planning an imminent attack on a police station. Driving such a slow and cumbersome vehicle to the south, even along the labyrinth of unmanned country lanes that cross the border invited detection from army and police helicopter surveillance. The RUC's best bet was to prepare for an attack similar to two others that had recently used hijacked bulldozers to ram gates. There were a limited number of targets in the area and most police stations were heavily guarded and protected with sound perimeter defences. There had already been twelve attacks on stations in Ulster in 1987 and the biggest had been mortared by IRA units from a safe distance. But security forces had countered the increasingly sophisticated and often accurate home-made mortars, constructed from steel tubes and mounted on flatbed lorries, by regularly patrolling open spaces where they might be parked to launch an attack. The disappearance of the JCB would confirm what was already suspected from other intelligence sources: that the East Tyrone brigade was planning the thirteenth raid on an RUC station close to home and had selected a lightly defended outpost. There were only a handful of likely targets in the area and all were already on alert.

It had been a simple matter to increase surveillance on the East Tyrone brigade quartermaster who, even before the Army Council had sanctioned the attack, had been identified to British intelligence as the man who would arm the service units with both bombs and weapons. It was even known who would carry it out and some of the ASU's known operatives who had been responsible for a mortar attack on Portadown RUC station three months earlier had been put under surveillance too.

The hiding place was pinpointed soon enough – a derelict farmhouse eight miles from Loughgall. The RUC's surveillance unit E4A, were watching when the bomb arrived, 300 pounds of explosive, ferried in component form to be packed into a barrel with a remote control device that would be activated as the terrorists escaped. The IRA cannot have known they were playing so easily into the security services' hands; they were leaving such a blatant trail of clues that the RUC must have wondered if the whole operation was a double-bluff.

But experience had taught them that the IRA's abilities differed widely between areas. Sophisticated tactical thinkers were in short supply in the border counties where the most common terrorist activities were attacks on unarmed, off-duty and often even retired security personnel in their own homes. So when the ASUs grouped for the attack at the derelict farmhouse and gunned the engines of the JCB and a blue Toyota delivery van stolen earlier in the day, the SAS were already waiting patiently.

Shortly before 7.20 p.m. on Friday 8 May, Loughgall was settling into its weekend routine. Wogan was on television, the church hall was packed with parents and youngsters for the annual gala of the Girls' Friendly Society and there was little traffic in the village high street. Inside the police station the SAS lay in wait.

They had been there since Wednesday when E4A had reported increased activity around the derelict barn. In fact the security forces had been alerted to a probable attack up to six weeks earlier. The theft of the JCB and the Toyota van that Friday from nearby Dungannon had merely confirmed the method and potential target for the attack, which was imminent. Lynagh was under observation south of the border in Monaghan, Kelly was being watched in Dungannon. As the SAS lay in wait the RUC kept up regular patrols from Loughgall station to maintain the pretence of normality. When shortly before 5 p.m. on Friday, Lynagh was observed crossing the border and Kelly left Dungannon, the security services knew they would not have to wait much longer. E4A, keeping its vigil at the derelict barn watched the terrorists arrive. All eight men donned blue boiler suits, plimsolls and balaclavas – the off-the-shelf combat uniform of the IRA. One climbed into the cab of the JCB and the remaining seven terrorists clambered into the Toyota. It might have been wise to have chosen a vehicle with side windows they could fire through if under attack. But the Toyota was a simple delivery van.

Only the two men on the front seats could bring their weapons to bear if stopped, the other five were locked in the rear.

Armed to the teeth and expecting little or no resistance, the terrorists drove to Loughgall. Just outside the village the JCB stopped and the Toyota drove into the centre to reconnoitre the station. All seemed quiet, there were perhaps no more than five officers on duty inside; the surprise attack would be over in seconds, the bomb detonated before the officers had time to draw their weapons.

But the stolen Toyota had been immediately identified within the station and as it drove slowly back to rendezvous with the JCB the SAS slipped silently into pre-determined positions in adjacent hedgerows and within the station's 'anti-rocket' wire mesh fence. Two soldiers were stationed further along the road to cut off the retreat and members of the RUCs stood ready to set up road blocks should the terrorists escape.

Shortly after 7.20 p.m., the Toyota appeared in front of the police station, jerked to a stop and men inside piled out. Then, like something out of *Gunfight at the OK Corrall*, they all lined up and began blazing away at the police station with a mixture of rifle and machine-gun fire. This mad act of bravado was supposed to give cover to the JCB, which came lumbering into the square, veered right, crashed through the perimeter fence and came to a halt in front of the station.

No one knows what came first, the blast from the 300-pound bomb or the SAS opening fire. Parents and children in the church hall believe the gunfire came first followed by the bomb blast. They dived for the floor as bullets ricocheted off walls, and estimated the gun battle lasted ten minutes. More likely it was all over in seconds. The explosion was devastating; half the station, which had been evacuated as terrorists prepared their final run in, was flattened to the ground. Timber, bricks and roof tiles rained down on the village and the JCB was lifted like a feather on the wind and dumped back in the middle of the road in a heap of scarred yellow metal, twisted beyond recognition.

To the terrorists, who clearly thought they were out for an evening's entertainment, the sudden assault by the SAS men came as a rude shock. The gunman at the controls of the JCB was able to clamber down from his vehicle and, with his two comrades, run back towards the Toyota van parked opposite. Others of the group caught in the open sought the feeble sanctuary of the van and scrabbled to get inside. But the bullets cut through the thin shell of the van's body and all the terrorists were killed.

In the heat of the moment, the cover squad, hiding in a small bluebell wood further down the road, opened fire on a white Citroën that was racing from the scene. The SAS thought a third IRA vehicle was making its escape and opened fire. Inside, innocent motorist Anthony Hughes died at the wheel and his brother, Oliver, was hit in the head and chest. As soon as the bullets had begun to fly, the RUC had moved into action cutting off the village from the outside world in case any terrorists might survive and escape by car. Road blocks were set up as far as twenty miles away to search for any terrorists returning to Tyrone from Armagh along the M1 motorway and A29. Ambulances took critically injured Oliver Hughes, two policemen and a soldier, wounded in the explosion, to Craigavon hospital in Portadown.

Within minutes Gazelle army helicopters hovered overhead, their searchlights scanning the surrounding countryside in the twilight in case IRA observers had been stationed near the village to cover the gunmen's retreat. Another dropped down in the playing field opposite the station, and the SAS men, always quick to leave the scene of an ambush to avoid identification or press photographers, clambered aboard to be flown back to their base.

That weekend newspaper headlines trumpeted the SAS victory. AMBUSHED announced the *Daily Mail*, IRA TOP GUNS WIPED OUT splashed the *News of the World*. The IRA had suffered its worst defeat since the 1920s; the entire East Tyrone brigade and two of the most dangerous gunmen on active service had been annihilated. In Ireland the responsibility for the death of the innocent motorist Anthony Hughes was placed squarely at the feet of the IRA's top brass by the republic's foreign minister Brian Lenihan, while Ulster unionist politicians predictably praised the SAS operation and called for more. Republicans quickly countered with the ridiculous notion that all eight members of the active service units had been executed with single shots to the back of their heads from close range. Quite how the SAS had managed to approach eight heavily armed terrorists, persuade them to hand over their weapons, line them up, force them to their knees and neck-shoot them in turn, was not explained, nor were the eye witness reports of the firefight. In Belfast, where black flags lined the Falls Road, Gerry Adams was philosophical as he addressed 2,000 republicans in an attempt to restore their shattered morale: 'I believe that the IRA volunteers would have understood the risk they were taking.' His close aide, Danny Morrison, attempted defiance, hinting darkly that revenge would be swift and merciless and that Mrs Thatcher herself would pay.

That night youths rioted in West Belfast and Londonderry but the petrol bombing of RUC patrols was not sustained and the street gangs were quickly dispersed. A huge body blow had been delivered to the IRA movement and attempts to gain sympathy for the gunmen were quickly negated when the RUC put their weaponry on show at Gough barracks in Armagh. The terrorists had been well-armed; the eight guns recovered from the scene included three Heckler and Koch rifles, two Belgian FNC automatic rifles (the standard Nato infantry weapon until

115

two years ago) and a Ruger Magnum revolver with the firepower to blow a hole in a brick wall. What served to dilute any propaganda the IRA hoped to gain from the ambush and their subsequent accusations of an SAS shoot-to-kill policy were the ballistic reports on the weapons recovered.

The guns had been used in at least seven assassinations in Ulster in the previous sixteen months and in another ten attempted murders – the victims had all been members of the security forces or building workers and tradesmen working for them. The Ruger had been used in the murder of part-time UDR soldier Thomas Irwin in Omagh in County Tyrone in March 1986. Three months later it was used to gun down Protestant business man John Kyle as he drank in his local pub in Greencastle in the same county. In October the Ruger had been issued again by IRA quartermasters to take the life of shopowner Kenneth Johnstone in Magherafelt in County Londonderry.

One of the Heckler and Koch rifles was used to kill UDR soldier Martin Blaney in the same month and, shortly before the Loughgall ambush, to murder UDR major George Shaw. It was Paddy Kelly's personal weapon and both men had died in the gunman's home town, Dungannon. Two other Heckler and Kochs were used to kill UDR Private William Graham at his farm in Pomeroy in County Tyrone in April 1987, and Protestant Harry Henry in Moneymore, County Londonderry. The FNC rifles had also been used in the killing of Graham and Henry. The weapons, returned to IRA quartermasters and their hiding places, had been used in unsuccessful murder attempts in three other counties.

The IRA made another feeble attempt to salvage some credibility from the disastrous operation; it announced that several active servicemen had managed to escape the SAS ambush and had been taken to 'a safe haven'. The *Republican News* announced that one survivor would come forward soon to give his account of how the SAS had delivered the *coup de grâce* to his comrades, but it was an empty promise. The anonymous man was never produced.

In the aftermath speculation was rife. Newspaper reports quoting the usually unnamed intelligence sources, 'revealed' that a high ranking 'sleeper', a mole in the IRA's ranks, had been activated to divulge forthcoming IRA plans that would allow the security forces to set up an ambush and score a quick victory to counter the swell of republican pride that had greeted news of Judge Gibson's assassination (see p.120). Certainly it was a subject high on the agenda of the IRA army council and GHQ staff which met to discuss its worst military defeat. The East Tyrone brigade had, under Kelly, become the most feared and ferocious terrorists outside Belfast. But it had also suffered setbacks against the security services in the eighteen months leading up to Loughgall – setbacks which suggested the security forces had a source inside.

In fact the ambush was handed to the SAS on a plate. The Tyrone brigade, had devised a plan so ambitious, yet so cumbersome, that routine surveillance and good detective work had given the RUC plenty of warning. The decision to join forces with Lynagh's Monaghan/Armagh brigade to mount a joint and, for the IRA, mass attack had led to leaks and operational mistakes, not least of which was the use of the JCB. It was inevitable that, as the gunmen prepared the bombing run, ferrying weapons, attending briefings and fetching explosives, routine surveillance on them would pay off. It was the arrival of Lynagh in the north, observed crossing the border from the south a few days before the attack for a final briefing with his co-conspirators, that alerted the security forces on the Wednesday before the raid, the day the SAS had moved quickly into the station. From that moment on it was a simple matter of waiting for the terrorists to assemble at the derelict farm where the bomb had been stored and the subsequent theft of the JCB. Waiting is, after all, what the SAS specialize in.

The IRA had driven into the trap with a cavalier disregard. Obviously they expected little resistance for they used a noisy vehicle for their approach, a van that could not provide them with firing points and a target that had only one approach road.

The IRA leadership quickly recognized the operation's shortcomings and dismissed thoughts of a British spy in their midst. Action was taken to ensure that the Loughgall debacle would not be repeated in the future but higher still on the agenda was the need to heal the widening gulf in the movement. There was a clamour for revenge on a large scale, hardliners were demanding action from the army council, or else they would take the matter into their own hands; the thirst for revenge would increase dramatically in the coming months of 1987.

CHAPTER 8

Stopping the Rot

The IRA's New Year message for 1987 was ritually optimistic. It promised 'tangible success in the war of national liberation' but delivered only political and military mayhem. Loughgall, Enniskillen and *Eksund* should have been landmarks in the pursuit of the republican cause – instead they stood as a grim reminder of the movement's fragmentation. A year that was to have seen giant leaps forward in the campaign brought unmitigated disaster.

In February, twenty-seven Sinn Fein candidates fought constituencies in the Irish general election called by Garret Fitzgerald. When the ballot papers were counted on 17 February, Sinn Fein was rocked by its dismal showing. It had received only 32,933 votes, a miserly 1.8 per cent share of the poll. Sinn Fein had fought an aggressive campaign but had proved itself hugely out of touch. To the electorate of Eire, it was a one-issue party and an irrelevant one at that; the question of re-unification generated little sympathy within a nation far more concerned with the appalling state of the economy, unemployment, taxation and a proud nation's slide into Third World status.

Their political ambitions trampled underfoot, even in the strongholds of Dublin and the border counties, the terrorists looked north again for a morale booster. The opportunity arose soon enough. The hardliners within the movement had already proved that Sinn Fein's political regression had not sapped its commitment to violence. In the first three months of the year nine Ulster policemen had been murdered and now the Northern Command seized an irresistible opportunity to score one of the 'tangible successes' predicted in the New Year.

Shortly after 8.30 a.m. on Saturday 25 April, Ulster's second most senior judge and his wife drove past a 500-pound IRA car bomb and died. Lord Justice Maurice Gibson was a natural target. Not only did he represent, at the highest level, British justice, but he had also cleared three RUC constables of the murder of three IRA men. They had, he summed up, brought the IRA men to the final court of justice. Judge Gibson, with his wife Cecily at the wheel, was returning from a holiday in Britain. Instead of travelling directly to Belfast, they had caught the ferry from Liverpool to Dun Laoghaire in the south. They were escorted north by the Garda, a customary precaution, and reached the border at 8.35 a.m. Until 1985 the RUC had met VIPs there to escort them home but since four officers had died on similar duty the practice had been suspended. The Gibsons drove on to the mile-long stretch of no man's land between north and south and, as they passed a parked car, an IRA active service unit triggered the bomb by remote control from a concealed position in the north near the border.

It was a huge propaganda success for the IRA and a security blunder for Anglo-Irish forces. For days after accusations of incompetence flew between security forces north and south but, as the dust settled, it became obvious that much of the blame rested with Judge Gibson himself for disobeying regular security precautions, failing to file details of his intended movements with RUC headquarters, for changing his travel plans and for dismissing the RUC guards he had been assigned.

The assassination of such a high ranking British official was celebrated late into the night in the IRA's illegal drinking dens that weekend; it was not only the removal of a hated enemy of the movement that gave them cause to rejoice but also the discord it had sewn. London and Dublin were squabbling again and the much vaunted co-operation entrenched in the Anglo-Irish accord seemed now little more than window dressing. February's election results were soon forgotten, but so too was April's 'tangible success'. For in County Armagh, just two weeks later, the SAS silenced the IRA's taunts of British incompetence. Eight hardliners, nearly a fifth of its entire frontline force and almost certainly the full complement of 'active' servicemen in the IRA's East Tyrone Brigade, were wiped out at Loughgall. The terrorists had suffered their biggest losses in a single action since 1921, throwing the movement and the army council into disarray.

The republican barons were confused; was it hi-tech British surveillance that had uncovered their plans to attack the barracks? Or were their own 'need to know' security precautions simply not enforced? Worse still, could it be the Brits had a high-ranking mole in their midst? Certainly East Tyrone's activities had been hampered by continued British successes; in recent years there had been a string of arrests and ambushes and IRA arms caches had been unearthed. The army council dismissed talk of a high-level informer; there had been suspicions, checks, searchers, but no traitor had been unmasked. Now it was the turn of the security forces and the British government to celebrate. The immediate reaction of the army council was to issue new standing orders; in future no operation requiring so many volunteers would be sanctioned, active service units would be given greater autonomy to plan and execute attacks, thus restricting knowledge of operational details. In Britain, Loughgall was hailed as a triumph for the security forces (even though an innocent motorist has been mistaken for an escaping terrorists and riddled with bullets). For Britain was heading for a general election and Mrs Thatcher for a landslide victory. As in Eire, earlier in the year, other issues were paramount.

For Gerry Adams and Sinn Fein, Loughgall promised a reversal of February's disastrous polling in the south; surely there would be a huge sympathy vote among Catholics voting in Ulster constituencies? It was a vain hope. On election day Sinn Fein's vote dived.

Support fell from the 13.5 per cent recorded in 1983 to little more than 11 per cent. Catholics switched to the legitimate republican alternative, the SDLP. It seemed twenty years of violence were beginning to sicken and exhaust all but the most committed.

The Anglo-Irish Accord had played its part. Catholics were beginning to believe London and Dublin could work together to ensure a long-term solution to the troubles; it had begun to eat away at the foundations of distrust and offered a significant alternative to bombs, bullets and burnt-out buildings. Perhaps Loughgall too had had its effect, for surely there could have been no greater illustration of the futility of violence.

So with the year barely half over, the republican cause had suffered severely on two fronts. Political defeats in elections north and south, its military humiliation at Loughgall; both gave the IRA/Sinn Fein propaganda machines cause to regret the smug optimism of the New Year. On 31 October, as Sinn Fein held its annual conference in Dublin, news came through that French customs had intercepted a small coaster off Brittany. In its hold were 1,000 Kalashnikov rifles, two tons of Czech-made Semtex plastic explosive, twenty ground-to-air missiles, mortars, rocket-propelled grenades that could rip open an armoured car, and enough ammunition to arm a small army. The crew were all Irish, the cargo was the gift of Colonel Gadaffi. Though it later transpired that four earlier shipments had reached their destination, the seizure still represented a huge loss to the IRA's arsenal – it seemed the cause could hardly fare much worse . . . until Enniskillen.

The bomb murdered innocent civilians, old men and young girls, all of them Protestant and all of them attending a simple ceremony, Remembrance Day. It was a blunder of unparalleled proportions. A day of respect and recognition honoured throughout the world, by girl guides, boy scouts, old soldiers and community servants had been blasted by a thirty-pound bomb – an untouchable target had been violated. The Protestant town in largely Catholic Fermanagh, buried eleven later that week; the secondary school headmaster was left a paralysed vegetable in his hospital bed and more than sixty others were seriously injured. One death in particular had touched the nation and, indeed, the world beyond. Marie Wilson, just twenty years old had been a nurse at the Royal Victoria Hospital in the republican stronghold around the Falls Road. She had travelled home to Enniskillen for a weekend with her parents and had walked to the cenotaph in the town centre that Sunday morning to watch the simple ceremony with her father Gordon. It was his heartrending description of their last moments together as they lay trapped under the rubble, she reaching out to hold his hand before dying from her injuries, that spanned the religious divide. And later, when he appeared on television to forgive the men who had murdered his beloved daughter, that moment of rare Christian warmth touched hearts, both Catholic and Protestant.

In Dublin, thousands signed a book of condolence; in Moscow, Tass denounced the 'barbaric' murders. There had been shock and revulsion even among the staunchly republican Irish community in New York.

The bombers had naïvely expected their attacks would

be celebrated as a victory for the cause, retribution for the massacre at Loughgall. Instead they had given the enemy an enormous opportunity to reinforce its charge that the so-called freedom fighters were nothing more than murderous thugs. The significance of the blunder was not lost on Gerry Adams; always ready to twist an argument to the IRA's advantage, this time Adams didn't even bother to duck and weave the invitations from foreign journalists to condemn the outrage. He publically expressed his regret for the bombing and offered his sympathies to the families of the victims. It was the closest Adams has ever come to condemning the actions of his own, indeed in an interview later he was to promise there would be no more Enniskillens: 'I think the IRA, in accepting responsibility . . . has signalled that they are going to ensure that there are no more Enniskillens.'

For the IRA army council had taken the unprecedented step of issuing a statement expressing 'deep regret'. It attempted to lay the blame on British security men insisting that routine scanning of high frequency airwaves had triggered the bomb's remote control device, but it was an empty attempt to apportion blame; the bomb had been armed with a timing device. The statement ended with an equally empty apology: 'Nothing we can say . . . will compensate the feelings of the injured or bereaved.'

The root cause of the blunder was historical. IRA cells in the border strongholds have always enjoyed a level of autonomy from the politically astute masters in Belfast and Dublin. Generally they were given freedom to choose their targets and operate independently, largely because too much control from above caused grass-roots resentment and dissent that could be both disruptive and dangerous in such a tightly knit organization. The masters exercised what little control they could by 'vetting' only the more ambitious plans of these country units that required logistical back-up from other ASUs. That control had been relaxed, however, in the light of Loughgall. Fear of infiltration had made the IRA much more secretive, and the ASUs had been given freedom to operate independently. Freed from the constraints of the leadership, the politically unsophisticated ASU had selected its target without thought to the damage such an atrocity might do to the republican image. The IRA had never looked less like freedom fighters nor so much like blood-crazed psychopaths.

It was a devastating own goal. Murdering policemen, soldiers and judges could always be passed off as the action of a force struggling against an occupying power. Who now would find sympathy with a cause that murdered innocents in the act of honouring their dead?

It was a bitter blow to Adams in particular. The 1980s had seen his influence in the republican movement grow to the point where he had finally, after a ten year battle, wrested control of Sinn Fein and the IRA from the old hands south of the border; he had installed his own hand-picked men in positions of authority and he had rallied support to the new republican ideal; a struggle incorporating the bullet *and* the ballot box. But it had been a long time coming.

Adams and his supporters had grown up with the troubles. Intelligent, quick witted and politically sophisticated, they had watched a moribund organization revive itself during the late sixties and early seventies to fight first the sectarian violence unleashed upon the Catholic minority in Ulster and then the British forces sent in to protect them. Adams had occupied a ringside seat in the early days of the troubles. At just twenty-one, in 1971, he was already the commanding officer of the 'second Battalion, Belfast' and was recognized as second in command of the Belfast Brigade. He was already one of the IRA élite – no further proof was necessary than his release from internment to fly from Belfast's Aldergrove airport in July 1972 to meet the newly created minister for Northern Ireland, Willie Whitelaw. The IRA, struggling for support amongst the Catholic community, had called a truce and requested talks. Whitelaw had agreed and Adams was one of the six leaders who met him for the ill-fated talks in the Chelsea home of Paul Channon, now Transport Minister.

Adams and his followers had watched the IRA stagnate in the 1970s, bombing, murdering and kneecapping in the north to little effect. It seemed the leadership had no clear strategy that might take it anywhere near the ultimate goal of a united Ireland; indeed the IRA had locked itself into the romantic idealism of armed struggle – chasing its tail to a standstill. The political masters in the south had little control over the hard men doing battle in the north; the movement needed cohesion and control but above all it needed clarity of vision.

Adams was interned in 1973, jailed for trying to escape, and released in 1976. This time internment in Long Kesh had interrupted Adams' relentless progress through the terrorist ranks but on his release he quickly resumed his quest for power gaining election to the IRA's Army Council. He had built upon his unrivalled reputation for brilliant tactical thinking and leadership by sharpening his intellectual grasp of republicanism and revolution among the older and wiser heads interned with him. He had left Long Kesh convinced that he would provide the IRA with the clarity of vision it required; that vision was political struggle enhanced by the selective use of violence on 'legitimate' targets of the occupying forces.

It quickly became apparent that Adams was the man to lead the republican movement into the eighties. He had already orchestrated his father's frustrated dream of a 'Northern Command', operating independently on the political and military masters who ran operations from the south. He had persuaded the republican barons that the northern units, who bore the brunt of the fighting, were operating at the far end of a long, inefficient leash. Permission was given for an autonomous command structure in Ulster in autumn 1976. Adams was immediately elected to the seven-man Army Council, the supreme ruling body of the political and military arms of republicanism. It was the beginning of the end for the terrorist controllers who had for so long operated from the relative safety of the south. By 1979 Adams had assumed effective control of the Ulster ASUs and the movement itself; the Army Council, now predominantly Ulstermen and elected by IRA delegates, chose him to be their chief of staff.

Together with two other young and determined northerners – Danny Morrison and Martin McGuiness – Adams set about orchestrating the IRA's activities to maximize impact on Whitehall. Out of the 'dirty protests,' in which republican prisoners discarded clothes and smeared excrement on their cells, grew the hunger strikes which were to give Adams and his supporters their chance to redefine their movement's strategy.

Until the death of a little-known MP in March 1981, Sinn Fein's attitude to elections in the north had remained ambiguous throughout the troubles; to participate not only suggested acceptance of the British system in Ulster but also carried the unacceptable liability of exposure. Defeat in elections would hardly sustain the movement's claim to represent the people of Ireland, north or south. But when Frank Maguire, an independent republican MP for Fermanagh and South Tyrone, died, hunger strikers in the Maze prison grasped the opportunity to regenerate their protest which had failed in its aim of igniting the public interest and thus winning support for its objective – political prisoner status. The first IRA man to start the hunger strike, Bobby Sands, decided that standing as a candidate in the by-election would light the blue touchpaper that had so far refused to burn.

On 9 April, Sands polled a staggering 30,000 votes to

win a seat in the parliament he could not and would never occupy. His election triumph had ignited not only Ireland and Britain but the world too; however, it was in vain. Mrs Thatcher's government stood firm; IRA prisoners were common criminals still to be denied political status. Sands died on the sixty-fifth day of his hunger strike, less than a month after winning the by-election. Few could deny the courage Sands had shown and more than 100,000 attended his funeral. It had been a stupendous propaganda triumph for Sinn Fein and the IRA; more importantly for Adams, now both chief of staff of the Army Council and vice-president of Sinn Fein, it had given him a firm foundation to rally the movement behind his firm belief that Sinn Fein must offer more than violence to further the cause.

Sands' support had convinced Adams that the time was now right to present Sinn Fein as an alternative political force in both the north and the south, and that Sands' death had acted as a catalyst to attract electoral support that would sustain its growth. At Sinn Fein's annual conference in Dublin in late 1981, Adams received the near unanimous endorsement of his followers; the republicans had, for the first time, voted to pursue a policy of electioneering. The new strategy was summed up in an enduring speech by Adams' then supporter and now rival Danny Morrison.

'Will anyone here object if, with a ballot paper in this hand and an Armalite in this hand, we take power in Ireland?'

Adams' hold on the movement was now formally complete. The commanders in the south were isolated and left with a backseat role to provide training, funds, safe houses and logistical support for the leaders and their troops in the north. Moreover, Sinn Fein, which Adams himself had always described as the IRA's underlining, had finally found a role for itself. Soon he was elected president and relinguished the post of chief of staff of the Army Council; he no longer needed the job with his own men in the majority, and set about making the fighting men subservient to the party he presided over. He now had control of both arms of the republican machine.

It was a masterstroke only as long as Adams could deliver the 'tangible successes' the two-pronged strategy sought. Entering the political arena, opening advice centres in republican areas of the north, installing the paraphernalia of party machinery inevitably required a reallocation of resources and it was the fighting units, the ASUs and their brigade staffs that suffered. As Adams pursued his political ambitions he was careful to avoid a split in the ranks, giving the hardline men of violence their head every now and then to plan audacious and bloody attacks. The campaign, now directed at Britain and 'legitimate targets' scored notable successes and failures; the Harrods bombing in 1983, a blatant attack on innocent shoppers, was widely condemned but served a purpose in carrying the Ulster troubles on to the streets of London. The Brighton bombing a year later demonstrated a new capability to strike at the very heart of the establishment. But military success was not matched by political progress. Sinn Fein had firmly entrenched itself in the north on local councils but by the mid-eighties, the party was shedding votes as the 1984 Anglo-Irish Accord, giving Dublin a say in the administration of Ulster, gave Catholics new hope of a peaceful solution. Adams now sought to muscle-up the Sinn Fein party machinery in the south, as he had done in the north, arguing that Dublin's signature on the Hillsborough Accord had to be erased by political gains in Eire. The agreement had been widely welcomed in the south, the Garda was successfully allying itself to the RUC and British troops in cross border operations. What was the use of fighting for a united Ireland on only one side of the border?

The hardliners who had watched their resources dwindle to finance political activities envisaged yet more cuts in their funding and began to make waves. Throughout 1986 and 1987, support for the hawks within republican ranks had been dwindling, and each military defeat took its toll on morale. The twin strategy of the 'bullet and the ballot box', using military action to wear down British resolve while demonstrating the depths of support for Sinn Fein and the IRA by increasing its electoral vote, was beginning to unravel and there were signs that the movement might fragment. Adams' political evolution had seen some victories. Choosing his moment carefully he had scored some notable and subtle successes over the hardliners, even, on one occasion, persuading the army council to court-martial four particularly dangerous and rebellious activists. But each new slide in the polls diminished Adams' powerbase. He still enjoyed effective control but now even his own Belfast supporters were beginning to believe the armed struggle had suffered for the sake of only limited political success.

By 1987 the IRA had lost a sizeable chunk of its fighting force to security service successes and Adams' restructuring. From a hard core of 1,000 IRA activists in 1980, the 'army' had dwindled to less than 250 of which barely 50 formed the Active Service Unit strength. The rest provided technical and logistical support, and a decreasing pool of 'reserves'. Security service success in 1986 had cut the ASU strength to 40, and Loughgall and other incidents in the first half of 1987 had further reduced frontline strength to perhaps no more than 30 men with 150 reserves. Recruitment was tailing off, there was talk of ultimate defeat. The Belfast brigade was facing a near mutiny for a large section was threatening to stand down unless it was allowed to increase the tempo of its military campaign. It was a serious crisis and the leadership acceded to the hardliners' demands. They would be allowed their revenge for Loughgall which itself had been designed to avenge earlier defeats.

The creation of a Special Active Service Unit, drawn from the ranks of the IRA's most accomplished and seasoned fighters, was sanctioned. Such units had been used at random in the past. Working directly to the Army Council, their duties would be to prepare and carry out a major strategic operation, the kind that required sophisticated long-term planning, costly logistical back-up and high-calibre personnel. The hardliners would get their revenge and send a signal to the movement that the downward spiral had been halted. It was already well advanced with its plans when the Enniskillen bloodbath hit the headlines. Now, more than ever, the republican cause needed a military success of Brighton proportions – and Adams needed it to dampen the flaring tempers of his hardliners if he was to keep his grip.

CHAPTER 9

Revenge

They met rarely and in secret, their intentions were known only to the Army Council and its GHQ staff, and their orders were simple: revenge. Since the crisis meeting in a west Belfast safe house, shortly after Loughgall, unprecedented precautions had been taken to ensure this time there would be no leaks, no clues for the security forces to detect and unravel. The Army Council inquest had been long and argumentative, recriminations had flowed thick and fast and there was an 'open verdict'; no one really knew if the movement had been betrayed by careless talk or a traitor. It was best, for morale's sake, to err in favour of a loose tongue. Once concluded, however, the inquest moved quickly on to the next item on the agenda. The decision to revenge the deaths of eight colleagues and rock the enemy on its heels was taken quickly and with little debate. The cause was in desperate need of a military success that would raise and rejuvenate the tattered standard of republicanism, galvanize the depleted and demoralized 'army' and blast the smile of smug satisfaction off the face of the British establishment.

The target would have to be carefully selected. A soft option inside Northern Ireland was out of the question. Loughgall had shown the risks involved; the RUC, E4A and the SAS had the odds stacked heavily in their favour.

Besides, where in Ulster was there a target that could rank high enough to rock the British on their heels, that was not already too well fortified and guarded? The mainland offered more choice but after Harrods, Hyde Park and Brighton, the police, MI 5 and the British Army had increased security. They were vigilant and prepared and the IRA could not risk losing more personnel. It would have to be abroad. The network of republican supporters among the ex-patriate communities of Holland, Belgium and West Germany could offer logistical support, safe houses and personnel, but IRA operations against British bases were once again limited by tight security; hit-and-run attacks on off-duty servicemen had been all the IRA could safely manage, and the murder of a handful of nightclubbing soldiers or an officer on leave would hardly rattle the British government to its core.

Political assassinations were a possibility but again, after Mountbatten and Brighton, worthwhile targets were heavily guarded. What was needed was a target where the Union Jack flew proudly but peacefully, where security lulled by routine had grown unintentionally lax, where the British Army wasn't always looking over its shoulder (for terrorists or the Warsaw Pact). Gibraltar was an obvious choice.

There can be few places in the world, as any atlas will demonstrate, where Union Jacks are so cherished that they decorate the walls of houses as republican murals decorate the streets of west Belfast. There can be few corners left of the old empire where the British Army remains a welcome presence, or a British administration preferred above all else or British culture is nurtured so protectively. Gibraltar is perhaps the only corner left where the locals refuse to part with their past so adamantly that their telephone boxes and bobbies are modelled on their British counterparts. No more than a rock, towering at the tip of southern Spain over the narrow straits where the Mediterranean meets the Atlantic, Gibraltar was the perfect target – a Lisburn or Catterick in the sun.

For the IRA, Gibraltar was ripe. It was the regular R & R destination of regiments completing tours of duty in Ulster – they would serve a tour on the streets of Belfast and then retire for a tour on the rock where duties were light, and locals friendly – as near as the military could get to a holiday camp. Its connections with the troubles of Ulster extended beyond its recreational role; inside the rock itself, a mass of tunnels and caverns excavated during the World Wars housed engineering workshops, Nato arsenals and a life-sized, prefabricated replica of a west Belfast neighbourhood. Complete with republican murals, terraced houses and burnt-out barricades of 'hijacked' cars, a resident regiment would sharpen up its patrol skills here before returning to Northern Ireland. As a military outpost its role is largely confined to 'ears and eyes' duty for Nato, watching Russian naval movements through the straits and listening to its communications traffic. It is a vital link in the worldwide intelligence-gathering network of GCHQ but the covert operations hide behind a cloak of largely anachronistic and ceremonial duties to celebrate the British military presence that is symbolized by the Trafalgar cemetery. Outside Buckingham Palace, Horseguards Parade and the Edinburgh Tattoo there can be few locations where British imperialism and military pomp are so regularly paraded at its 'busby, buckle and brass' best. An attack on a military target in Gibraltar was the next best thing to blowing up the Household Cavalry or Sandhurst, and would, decided the Army Council of the IRA, be a damned sight less complicated. It would shock Whitehall and astound Britain. Moreover, it would surprise a world long used to bomb blasts in Ulster or London; such an attack could not be confined to the small print of the world's press and the blazoned headlines would sorely embarrass the enemy. It was a special job and it required special treatment and special treatment required a special team.

Shattered by losses and lacking experienced personnel, the IRA had been wary of calling on its battle-scarred active service units for obvious reasons. If there was a mole, they risked betrayal; the Northern Command's campaign against the security forces had to be maintained and, besides, most of the gunmen and bombers were local men, with local knowledge and would be out of their depth abroad. The cells in Holland and Belgium were stretched too tightly planning a campaign against servicemen in West Germany. When circumstances dictated, the Council had, in the past, formed a Special Active Service Unit which worked directly to the GHQ staff on overly ambitious objectives and it did so now. Although depleted by the success of the security forces, the IRA's ranks had been swollen slightly in the mid-eighties by active service members freed from prison after serving jail sentences for terrorist offences in the heavy campaigning of the early seventies. It was policy to keep convicted terrorists at arm's length on their release, assigning them to political tasks within Sinn Fein for a year or so. Two reasons governed the rehabilitation of released detainees; the movement could not welcome back into frontline operations a member who might have been 'turned' in prison. The other consideration was the newly freed member's commitment and stability. The desire for revenge after a long and uncomfortable sentence might make a volunteer unstable and unreliable thus threatening the safety of his fellows. Equally, his commitment to the cause, initially rekindled by freedom, could be diluted by domestic pressures after a few months of freedom. Ever present also was the likelihood of surveillance, for it would be only common sense for the security forces to keep watch on a released prisoner's movements. The threat of revoked parole and a return to prison might persuade a volunteer that collaboration was preferable to confinement. Nevertheless, it was from this small core of former activists and sidelined soldiers that the Army Council now instructed the GHQ to assemble the special active service unit.

With a limited choice, it was an easy process. Three committed terrorists stood out. One had been released from prison a year earlier and was highly regarded in the movement for intellect and commitment, the other was a hardline veteran of violence who had been expelled from the movement for his hawkish belief in no-holds-barred bloodshed. The third was young, technically adept with explosives and had no police record. Such a team not actively involved in terrorist operations in Ulster might escape the regular scrutiny of the security forces. Mairead Farrell was the young idealist put in charge. Perhaps the highest-ranking woman in the IRA, she was already a political and military veteran of the troubles at just thirty-one and had served ten years in prison, commanding the IRA women in Armagh jail and leading the hunger strike with an eigthteen-day fast herself. She was an Adams acolyte, tough, resourceful and a natural leader; like Adams she believed the ballot box was the only way the republican movement could ever unite Ireland and like Adams she looked upon violence as a legitimate weapon to wield until British resolve cracked or the electorate fell in behind Sinn Fein. She was also the most unlikely IRA recruit.

The IRA can claim few middle-class, ex-convent schoolgirls as converts to the cause. Its support has always been largely drawn from the backstreets; from families with long republican traditions, with interned brothers, fathers, uncles or cousins. Mairead Farrell was an exception. As a schoolgirl at Rathmore Grammar she had been bright, vivacious and untouched by the violence despite growing up in the Catholics Falls Road and Andersonstown area. Her family voted SDLP and her only link with the IRA had been her grandfather's internment in 1920 for refusing to drive a train for the British black and tans. Her closest shave with authority came when she laughed at a nun who fell, unhurt, downstairs and she preferred school plays to street politics. It was during her final months at school, studying hard for A levels, that she underwent her conversion.

Farrell liked to claim that it was the Bloody Sunday massacre in 1972 that turned her head to republicanism at the age of fifteen but it was in fact an adolescent crush. At Rathmore, studying for A levels, she suddenly developed a strong political fervour that put her at the forefront of dinner-table discussions at home. She had met Bobby Storey, a wild young IRA man whose charismatic idealism transformed her life (and his, for he was to be caught, convicted and sentenced to life imprisonment in the Maze). Soon she had dropped out of school and taken a job in an insurance office, studying IRA republicanism in preference to academic subjects. Unknown to her parents and largely apolitical brothers she was soon sucked into the maelstrom of violence. Less than a year after leaving school and just eighteen years old, Mairead Farrell found herself in a police cell under interrogation.

On Monday 5 April 1976, the IRA had launched a major offensive in Ulster in protest at the removal of political prisoner status for jailed IRA members. Such was the intensity of the campaign that resources had been stretched to the limit and the IRA had called on even its young supporters like Farrell to unleash the wave of bombings and shootings. Farrell had already been blooded in an assassination; indeed when her two colleagues complained they could not shoot the target because he was unarmed, she threatened to have them shot instead; naturally they chose to obey orders. It was on 5 April that she found herself with two more experienced male active servicemen, carrying Colt 45 pistols and suitcase bombs into the Conway Hotel in Dunmurry. They were to be placed in the dining room, the lounge and an upstairs hallway in the hope of killing the servicemen who frequented the hotel.

Farrell and her accomplices had held up the security men inside the hotel, planted their bombs and were running from it when Farrell heard a warning: 'Halt or I shoot.' She had lost contact with the two men, so she froze, dropping her Colt to the floor. Two hours later, under Special Branch interrogation, she heard her handiwork explode. Three days later, Farrell found herself on remand in Armagh jail. She was convicted and sentenced to fourteen years' imprisonment in just one day; three witnesses were called at her trial, she refused to recognize the court and sentence was handed down speedily. She put her time in Armagh to good use. It was natural that by the time she was released, she would have risen to become Officer Commanding the republican women inmates who followed her.

Protest was to be the central theme of Farrell's ten years in prison. It started as soon as she arrived: she was the first woman republican prisoner to be refused political status. Such status would give prisoners privileges over other criminal inmates; it could mean no prison work, personal clothing instead of uniforms, superior educational and recreational activities, more free time, more visits from family and friends. More importantly, if the British government were to give the IRA political status, it would acknowledge that there was a war in Northern Ireland, and by giving them the standing of an army in the field would imply an unacceptable legitimization of a terrorist group. Loss of that status was a bitter blow to republicans who, by the very nature of their crimes, faced long and tedious sentences. Farrell soon found herself an audience as more female republicans joined her in Armagh and they began their protest by refusing to do prison work. Instead Farrell organized classes in Irish, politics and republicanism but it wasn't until three years into her jail sentence that her name became familiar in the republican movement outside Armagh.

In February 1980, Farrell began the 'dirty protest'. Locked in their cells for twenty-three hours a day, two to a cell measuring no more than nine by six feet, the prisoners refused to slop out and began smearing their excreta on the cell walls and refusing to wash. The cell windows were boarded up to stop them emptying their chamber pots into the yard below, and the bare rooms became dark, filthy and germ-ridden. The protest lasted thirteen months, guards regularly hosed down the walls and the occupants and there was a brief respite, for a few weeks only, when Farrell began a hunger strike. Taking only water and salt tablets, she held out for eighteen days before calling it off.

This then was the republican heroine who emerged from prison in September 1986, Officer Commanding, Armagh Prison Women's Battalion. The day after her release she granted an interview to a magazine journalist. In between her protests she had studied politics and social sciences with the Open University and the young girl who had entered Armagh's gates ten years earlier had emerged a far more sophisticated and determined woman: 'I've learned a lot of self-discipline and the importance of strength and resolve,' she told *Magill*, the Dublin-based magazine. 'I'm going to continue my involvement with Sinn Fein, remain politically aware. Prison is an experience I'd never have got anywhere else.'

Her significance to the movement was highlighted shortly after her release. At Sinn Fein's annual conference in Dublin that autumn, Gerry Adams called her to the stage for a standing ovation. She threw herself into the movement, staffing Sinn Fein advice centres in Belfast, and administering to the needs of Catholics who had turned to such offices for help, be it social security problems, housing or a simple case of trouble with the neighbours. She was, reported *Magill*, a small, slender, pale-faced woman, calm and composed with no regrets. Farrell had begun the slow process of rehabilitation to the ranks of the IRA, her commitment to Sinn Fein a mere cover-story for her desire to become actively involved in terrorist operations once more.

When the decision to form a Special Active Service Unit to attack Gibraltar was taken nine months after her release, Farrell had served her time in the Falls Road Women's Centre. She had the cool, calculating resolve to see an attack through and the risk of civilian casualties did nothing to diminish her commitment. For Adams, who knew the damage that civilian casualties caused the movement, she was a perfect choice; she supported his stance and would ensure maximum damage was visited upon British forces while minimizing the threat to civilians. Tough, ruthless but politically sophisticated Farrell was the perfect foil for the second member of the team, the hard man from the Lower Falls area of Belfast.

Danny McCann was the hawk Adams had expelled from the movement in 1985 when the more bloodthirsty gunmen in the ranks, angered by the demotion of military activity in favour of political advances, had threatened to turn renegade and initiate a campaign of murder and mayhem that would hamper the Adams strategy. Both men distrusted each other and would have gladly seen the other fall from grace, but McCann was a powerful force in Belfast, with many supporters. He had won reinstatement and now the active service list was in need of reinforcement. His record was second to none. With a string of successful operations to his credit and newly returned to the ranks he was selected for his operational expertise.

Though only thirty and married with two children, McCann had been the mastermind behind many IRA bombings and assassinations in Belfast for a number of years. He was a butcher by trade and a butcher by reputation, regarded by the RUC as the most important IRA terrorist in the city. Unlike Farrell, he had entered the IRA as a street urchin, dropping out of school to stone British soldiers and firebomb RUC patrols. Such was his profile in the movement that loyalists had tried to murder him, bursting into his home with shotguns once when he was out. On another occasion a cross bearing his name had been pinned to his front door. True to the cause he had immediately blamed the security forces for the attempt on his life, while failing to explain why he was still alive if the RUC wanted him dead. McCann had largely defeated attempts to bring him to justice. He was jailed for two years in 1979 for possessing a detonator and joined the dirty protest; he was arrested several times in 1981 and 1982 but was released and in 1983 he was charged with possession of firearms on the evidence of a supergrass and cleared. It was during periods on remand that he had married his wife Margaret and fathered two sons, Daniel and Meidhb.

He was an elusive gunman; Special Branch believed him to be a member of the RPG-7 gang which was responsible for a number of botched rocket-propelled grenade attacks, including one in which the wheel of a coal lorry was the only thing damaged. It was in the mid 1980s however that he had refined his art; RUC files linked him to twenty-six murders including the deaths of two undercover special branch officers on duty at Belfast docks.

Blonde haired and thick set, McCann was considered the IRA expert in close-quarter killing and thus would provide the cutting edge in the Special Active Service Unit. In fact, even while he was helping set up the Gibraltar operation, McCann carried out a number of terrorist attacks in Belfast. As one police source puts it: 'Towards the end he just became addicted to the violence.' With Farrell's brains and McCann's brawn, the team needed two more recruits.

Sean Savage was only twenty-three but he had the technical experience that the team required. His bomb-making had already come to the attention of the security forces who can tell the identity of a bomb-builder from the method of construction and the materials used. A length of wire cut to a certain length or twisted in a certain direction around a terminal was as good as a fingerprint; every terrorist bomb-builder had his hallmark and Savage was no exception. He had left school with eight O levels and had joined the IRA at the age of seventeen, his academic qualifications marking him out for technical duties. He was a keen cyclist, Gaelic footballer, amateur photographer and Irish language enthusiast, like Farrell; a cut above the usual recruit but equally impressed by the savagery of McCann.

When the team met for the first time with the GHQ staff, they were joined by a fourth member of the team, a woman selected for her expertise with disguises who would travel ahead, gathering intelligence, reconnoitring the target area and advising on routes of approach and escape. She was to be the gang's eyes and ears, reporting back from Gibraltar on the optimum target and the optimum time to strike. In the weeks that followed Loughgall she visited Gibraltar several times using the alias Mary Parkin.

An attack on Gibraltar would have to be carefully planned; there were many civilians on the rock, barracks were carefully guarded, the naval docks and the arsenal inside the rock regularly patrolled. But Gibraltar town itself offered numerous military targets, the Governor's residence, the courts, officers' and NCOs' messes among them. But one event in particular offered the best chance of a successful operation, a target that could be relied on to present itself with military precision and regularity: the changing of the guard.

Every Tuesday morning the band of the Royal Anglian Regiment, which had just arrived in Gibraltar after a tour in Northern Ireland, paraded up and down the main street as the guard outside the Governor's residence was changed. Nearly seventy bandsmen in ceremonial dress would assemble in a small square off the main thoroughfare, march for twenty mintues playing regimental favourites before marching out of sight to their dispersal point, a seemingly quiet square where a car bomb could be parked without attracting too much attention and apparently well away from the main tourist streets of Gibraltar town. The IRA had its team, its target and its chance to redress the balance.

CHAPTER 10

Gibraltar

Everyone knew he was coming, but no one knew quite when. In the end Sean Savage crossed the frontier some time after noon on Sunday 6 March, but Detective Constable Charles Huart of the Gibraltar Special Branch missed him. It was not until much later when he was spotted beside a parked white Renault 5, registration number M9149HW, near Inces Hall in the centre of town, that the group of soldiers, policemen and security experts, set up to combat the IRA plot to murder British military bandsmen in Gibraltar, knew that Operation Flavius was underway in earnest.

News that Sean Savage was in Spain reached London in the autumn of 1987. The routine interception of mail going to the homes of known IRA sympathizers in Ulster uncovered a postcard from Savage in the Costa del Sol telling his family that he was working abroad. It was something to note but at that stage not to be overly suspicious about. Savage had no criminal convictions, and although he had been charged in 1982 with being a member of the IRA and causing an explosion, the charges had been dropped. However his security file also carried the warning that he was suspected of being a bombmaker.

And there matters rested, until on 15 November two

men passed through Madrid's Barajas airport on their way from Malaga to Dublin. They both carried Irish passports, one in the name of Robert Wilfred Reilly and the other Brenden Coyne. In another of those routine checks that is the mark of modern day anti-terrorist work, Spanish immigration officials passed the details of the Irish passports to the Euro-terrorism experts in Madrid's Servicios de Información office. They in turn checked the passport details and photographs with London. Both passports were false and while the fact that Brenden Coyne was really Sean Savage was worrying enough, the identity of the second man set alarm bells ringing. It was Daniel Martin McCann, the butcher.

Although other intelligence information was available to the security forces, the combination of McCann and the suspected bomb maker Savage on the Costa del Sol could only mean either that the IRA had targeted someone among the 250,000 or so British residents on that stretch of coast, or, more likely, it was going for the soft underbelly of the British army in Europe – Gibraltar.

The Rock of Gibraltar is an unlovely place but to around 1,500 British army personnel it is home. It is some two miles long and bounded on three sides by the sea. The fourth is sealed by a high wire fence that is constantly patrolled. The changing of the guard outside the Governor's residence is a highpoint for tourists and was considered the terrorists' probable target. It is performed each Tuesday between 10.55 a.m. and 11.20 a.m. by the bandsmen of the resident regiment. It always attracts large crowds, but on 11 December last year the ceremony was suddenly cancelled forthwith.

The Governor's guardhouse needed refurbishing – urgently. This was true, but it also provided the security forces with much-needed time for further intelligence gathering and to prepare a welcoming committee for the suspected bombers. The offices of Gibraltar's Special

Branch are on the first floor of the colony's main police station. In the first week of January the six Special Branch officers were joined by a collection of plain-clothes officers with photographic and electronic surveillance equipment. It was assumed that a major drugs investigation was underway since the officers were often seen at vantage points in the town or loitering on park benches. No one chose to disabuse the local police.

However, time was running out. The changing of the guard ceremony was due to resume on 23 February. All the intelligence suggested that it would be the target, but no one could discover when. Then the surveillance work paid off. An Irish woman holidaying on the Costa del Sol had made two trips to the Rock on a Tuesday. Nothing surprising in that, except that she was travelling on a stolen passport in the name of Mary Parkin, the wife of a Press Association journalist; this then was the fourth member of the team, the expert in disguise. She was back on the Rock on 23 February to watch the first resumed changing of the guard ceremony. She in turn was watched by the MI5 surveillance team that had been shipped in earlier in the year. On the following Tuesday 1 March she was back again, and on each occasion she followed the route taken by the bandsmen of the Royal Anglian Regiment. The watching security officers concluded that she was reconnoitring for the arrival of an active service unit.

The information flowing back to London from Gibraltar was supplemented by fresh intelligence from Ulster and on Wednesday 2 March, the Joint Intelligence Committee, which reports directly to the prime minister, decided that the bombing was imminent. It alerted the Joint Operations Centre, which is manned by personnel from the Home and Foreign offices, the Ministry of Defence, the intelligence service, together with an SAS liaison officer. This group, based in the MOD, has the power to deploy the SAS and it now did so. A full special-projects team of sixteen men, which is on permanent standby at Stirling Lines barracks in Hereford, was ordered to Gibraltar.

The group included: a commander, who would later be identified only as soldier F, whose job it was to take overall command of the operation against the terrorists once the Gibraltar police had handed over their powers; his number two, soldier E, the tactical commander whose job it was to make sure the orders were carried out properly on the ground; and four further soldiers, A, B, C and D, who would, if all went well, arrest the three terrorists. Other men were available but were waiting elsewhere on the Rock, and were never used. The final member of the team was G, a bomb disposal expert who would act as adviser to both the civilian authorities as well as his colleagues in the SAS. The group flew out to Gibraltar on Thursday 3 March.

The operation was carefully set up so that when the terrorists were arrested and brought to court, the participation of both the SAS and the intelligence service would never be known. At every step of the operation in Gibraltar local Special Branch and other policemen worked alongside the SAS and MI5 (who were codenamed 'Snuffbox') so that, over time, a complete case could be established that would have stood up in court. When the bombers came to trial, it was planned that the Gibraltar police would take all the glory for cracking the case and the men from Britain would never even have been mentioned. But it was not to work out as planned and the result was the total exposure of the very people the operation had been designed to disguise – the SAS and MI5.

The intelligence service was now convinced that the bombing would take place on Tuesday 8 March, yet while the Joint Operations Centre was deliberating, there was still no sign that either McCann or Savage was getting ready to leave Belfast. Indeed both men had been seen outside a bar in the Falls Road on Wednesday 2 March. No matter, the authorities knew they were coming. They knew the third member and commander of the team was a formidable woman called Mairead Farrell, they knew what the target was, they were pretty certain that they knew the date, and they thought they knew exactly how the bombing would be carried out. So far the whole operation had been a staggering success; seldom before had the security forces been in such an enviable position. All they had to do now was watch, listen and wait. Unfortunately it didn't turn out quite like that.

The Cuerpo Nacional de Policia at Malaga's international airport had been briefed about the imminent arrival of the IRA active service unit, pictures of all three had been issued and they knew that Savage would almost certainly be travelling under the name Coyne, that McCann would be either Reilly or McArdle and that Farrell's false passport would probably carry the name Katherine Harper (née Smith). The chief inspector in charge of the operation had already set up a special control point. When flight IB657 arrived from Paris at 8.05 p.m. on Friday 4 March, the photographs were unnecessary. Savage and McCann came through immigration as Coyne and Reilly. McCann was carrying a small suitcase and a dark blue 'Puma' holdall, Savage carried a single black nylon travelling bag. Once through customs they made for the lavatories. The surveillance team held back since, in the words of the Spanish chief-inspector, both men looked around in 'a vigilant manner'. After a heart-stopping wait both men emerged and went to change money at the Banco Exterior de España, and there they were joined by a woman. Mairead Farrell had arrived, the unit was complete.

What happened next is a mystery. The official police

report into the incident makes sorry reading and offers no satisfactory explanation: 'Both men left the terminal and boarded a taxi, which it was not possible to follow, whilst the woman was lost from sight inside the building due to the number of people there at the time.' For British intelligence this was the low point of the operation. Despite their claims on Gibraltar, the Spanish police had been exceptionally co-operative in acting on information supplied by the British. But now, at the last moment and apparently through ineptitude, the whole operation was in jeopardy.

Both London and Gibraltar were informed immediately. The bombers were on the loose. The taxi driver was found and told the police that he had taken both men to the Hotel Florida in Fuengirola, which is some thirty kilometres down the coast from Malaga. The address of the hotel was written on a piece of paper which McCann had shown to the driver and he had left both of them on the steps of the hotel. There the trail went cold. Neither man was a guest at the hotel and no on had seen them there.

From that moment the Spanish police set about searching the hotels along the Costa del Sol. They were still showing photographs of McCann and Savage to hotel receptionists on the Sunday, some two hours after the three bombers had been shot dead in Gibraltar. In fact the two men had doubled-back and moved sixteen kilometres back up the coast to Torremolinos and the Hotel Residencia Escandinavia. The police were in confusion. Had the surveillance teams at Malaga been spotted? They need not have worried, the hotel switch was made solely because a 'good' terrorist always assumes the worst and covers his tracks.

The desk clerk at the Escandinavia was surprised to see the burly Irishman in a blue check suit asking for a room – it was past midnight. The man wanted a double room. His friend was waiting at a nearby hotel which, they had discovered, was full. Half an hour later the two men returned and checked into room 14 under the names of Coyne and McArdle. They paid 4,580 pesetas in advance for two nights. The room was at the back of the hotel with direct access to the street – a facility laid on for young men inclined to return late at night worse the wear for drink – and it had three beds. There now seems little doubt that Mairead Farrell slept there that Friday and Saturday night. When the police finally located the hotel they found make-up and women's clothes in the room and the hotel staff say that all three beds had been slept in, although they never saw a woman. The next morning Savage left the hotel and rented a white Renault 5 from Avis Car Rentals in Torremolinos. At almost the same time Mairead Farrell was on the telephone arranging to rent a second car from Marbella on the Sunday.

The two men had breakfast early on Sunday morning in a bar near the hotel, then together with Farrell they set off along the coast road towards Marbella in search of avenue Ricardo Sonano and the firm of Rent-A-Car Marbesol. This was the firm Farrell had telephoned on the previous day to check that it was open on a Sunday. Manuel Crespo, the rental clerk at Marbesol, remembers Farrell well – most of all the Irish accent, but also the white blouse, navy skirt and gaberdine mac. She paid the deposit in sterling and Crespo handed over a white, plastic key-ring and key for a white Ford Fiesta number MA 2732 AJ which was parked across the road opposite the office. At this time, Crespo is emphatic, Farrell was on her own. She left the office, got in the car, but then had difficulty getting it to start. Crespo went out and gave her a hand. Once the engine was running Farrell asked the way to Gibraltar and drove off. In fact she took the car no further than the basement area in the nearby Sun car park at Edificio Marbeland.

It is from the discovery after their deaths of the various cars hired by the bombers that their movements before they reached Gibraltar can be pieced together. At 5.00 p.m. on Sunday 6 March the Gibraltar police sent the following message to the police in La Linea, the Spanish town that butts on to the gates of Gibraltar: 'In a firearms accident at 15.30 hours three persons have been shot dead and they are suspected to be terrorists, it is felt that there could be a Ford Fiesta registration number MA 9317 AF parked in La Linea de la Concepción and that it could be packed with explosives'. The car, a red Fiesta, was discovered in the early hours of the following day, without explosives but in the glove compartment was a hotel key-ring on which was inscribed a large capital 'E', a car key for another Ford Fiesta registration number MA 2732 AJ and a rental contract for that car. It was these items that at last led the Spanish police to the hotel they had been searching for all over the weekend – the Escandinavia – and to the white Ford Fiesta packed with explosives in Marbella's Sun car park. The question that needed an answer was how did the car hired at 10 a.m. on Sunday acquire 141 pounds of Semtex when the woman who hired the car was in Gibraltar within a matter of hours? The link appears to have been the red Ford Fiesta found at La Linea.

On Thursday 3 March, just as the SAS were preparing to leave for Gibraltar and while Farrell, McCann and Savage were believed to be still in Belfast, an Irishman calling himself John Oakes booked into the A1 Andalus hotel in Torremolinos. Shortly before midday on Friday he went to a nearby car hire firm called Autoluis. He was in a hurry and needed a Ford Escort. There were no Escorts available, but he could have a Fiesta. That would have to do and the proprietor, Luis Cardon, recorded the details from his driving licence and passport. What stuck in Cardon's memory was Oakes' behaviour when signing the hire form. He started by using his left hand then switched to his right. As he wrote he seemed to be carefully copying the name Cardon had already entered at the top of the form, and even so he managed to write OAGES rather than Oakes and the writing was childish. Nonetheless Cardon didn't query the hiring. Oakes was respectably dressed in a grey suit and appeared well educated. Oakes then asked for directions to the nearest call box. He had to telephone his mother. On his return he surprised Cardon by asking where his girlfriend was. Cardon had seen no woman, but as Oakes drove off in the hire car there was a woman beside him with short, blonde hair. Mairead Farrell's hair was dark. To a Spaniard 'Mary Parkin's' hair would have been called blonde.

There was no other evidence to suggest that it was 'Parkin' with Oakes. It may have been Farrell. No one knows what time she arrived in Malaga. She certainly met McCann and Savage at the airport on the Friday, she was certainly seen at her parents' Belfast home on the Thursday in Belfast. It is possible but unlikely that she arrived on the same plane as McCann and Savage. Known terrorists try not to travel in groups. There is no shortage of scheduled flights into Malaga that would have allowed her to meet Oakes in Torremolinos by midday, but it would have been risky and not served any particular purpose. The identity of Oakes is also a mystery.

Another road to Gibraltar started from Belgium. In January 1988 the Belgian police raided a lock-up garage off the Rue Leon Lepage in Brussels. They were looking for clues to the Brabant killers, a gang of armed robbers who had murdered thirty people in a series of supermarket raids. Inside the garage was a battered Renault 4 with Dutch licence plates and inside the car were 110 pounds of Semtex explosives plus all the paraphernalia necessary to make a very powerful bomb indeed.

For many years Belgium had a reputation as a haven for terrorists. Members of the French terrorist group, Action Direct, had lived in Belgium while on the run from the French authorities and members of Germany's Red Army factions had found sanctuary in the country. The Belgian government took the view that as long as no acts of terrorism were carried out on their territory terrorism was not their responsibility. But in the early 1980s Belgium developed its own terrorist group, the CCC, or Fighting Communist Cells, which was responsible for almost thirty bombings between October 1984 and December 1985. The Belgians started to take terrorism seriously and formed their own counter-terrorism force and dramatically improved cooperation with their allies. The result was a routine sharing of intelligence and information on terrorists. New computerized data links now connect Western intelligence agencies and counter-terrorist forces and these have proved a successful addition to the armoury of the forces fighting terrorism. The success of these links was proved again this time.

Although the Belgian detectives were baffled by their find, they concluded in the end that they had stumbled on a terrorist's explosives cache. Police forces in Europe were informed in the search for answers. London believed it knew what was afoot. Long experience has taught explosives experts not only to be able to recognize the terrorist group that has made a particular bomb but also the bombmaker. They all have their individual ways of doing things. On this occasion London recognized the hand of the IRA and believed this was one more part of a giant conspiracy that would lead in the end to Gibraltar and the bandsmen of the Royal Anglian Regiment.

The authorities in London were not particularly shocked by the size of the Semtex cache found in

Brussels. They now knew that at least four coasters had reached Eire from Libya with guns and explosives before the arrest of the *Eksund*. Their guess was that something like three tonnes of Semtex had got through and it was to be expected that some of that would end up in Europe for use against British forces there. What worried them far more was the detonation system found in the battered Renault. It was a home-made radio control device operating on the VHF two-metre band. It was this piece of intelligence that coloured all the security service's thinking about the Gibraltar bomb and helped convince them that when the active service unit came it would be carrying a 'button job' that would allow the bomb to be detonated from some distance away and at exactly the right moment – no timing device mistakes as at Enniskillen when the bomb went off early.

The certainty that this was a 'button job' was drummed into the SAS men at their briefings. It was only after the shooting and the discovery of the Ford Fiesta in Marbella that the Spanish and Gibraltar police found out that the unit had decided, after all, to rely on a timer. However it was no ordinary timer.

The timing device found in the Marbella bomb was based on a design that had been perfected and used in more than 180 of the IRA's previous bombings. The design is based on the 'memo park', a small key-ring-sized timer generally used to warn motorists when their parking meter is running out or lorry drivers when they are in danger of exceeding their statutory driving period. The memo park had been modified so that the terrorist got up to an hour to set up a bomb before the timer automatically switched itself on. In effect it was a failsafe mechanism and two similar devices had been used in the Brighton bomb.

There remain two unanswered questions about the

Gibraltar affair: who chose the target and who provided the explosives? The target was very unusual in that it was outside the IRA's normal area of activity and it had clearly been marked for them many months before the bombing actually was planned. At the same time, other attacks carried out by the IRA in Europe in 1988 suggested that there had been good long-term surveillance carried out that had gathered very precise intelligence on the location and movement of British troops serving in Germany.

The British considered that a close friend of Danny McCann, Peter Anthony 'Pepe' Rooney, might well be the man responsible. Rooney, thirty-three, from James's Gardens, Belfast, had been jailed in 1981 for five years at Belfast Crown Court for possession of firearms, and the security forces believed that since his release he had renewed his ties with the IRA. For over a year, Rooney had been travelling Europe on a false passport and driving licence in in the name of James McCann. It is believed that he may have played a key role in scouting potential targets for IRA teams operating in Europe to provide some of the logistics back up for the Gibraltar ASU.

For the British, knowing the identity of the IRA intelligence officer in Europe was potentially a major coup. By following him, he could lead them to two other IRA units believed to be based in Europe, one in Holland and the other in England. Unfortunately, Rooney was arrested on 21 July 1988 as he was coming off the ferry from Brittany at Ringaskiddy, and charged with giving a false name and being a member of the IRA. The Irish policemen had no idea who they had caught and believed he was simply a low-level operator, but Rooney's arrest was a blow for the British.

The Irish eventually dropped the IRA charges but

Rooney was convicted of having a false passport and driving licence and sentenced to five months in jail. At his trial he issued a statement from the dock denying any involvement with the Gibraltar shooting.

There now seems little doubt that the Semtex for the Gibraltar bomb was collected in the red Ford Fiesta. This may have been done by a totally independent IRA team, whose faces and aliases were unknown to the authorities, flown into Spain for the purpose. The location of the explosive cache also remains a mystery. When the red Fiesta was found at La Linea it had covered 1,594 kilometres since being hired on the Friday. No one knows where it had been. What is certain is that on that Sunday morning the red Fiesta joined Farrell's white Fiesta in Marbella's Sun car park. The explosives, along with two six-volt batteries, four detonators, two timers, five pounds of Kalashnikov .762 ammunition, and a length of black wiring, were transferred from the red to the white Ford. This done, the active service unit then set off for the final act in their enterprise – Savage in the Renault 5 and McCann and Farrell in the red Fiesta, heading along the E26 for Gibraltar.

As far as Joseph Canepa, Gibraltar's police commissioner, was concerned everything that could be done had been done. His advice group was in place. Soldiers A, B, C and D from the SAS were on the streets, watching and waiting. Soldiers E, F and G from the SAS were in the operations room along with his own deputy Charles Colombo, Joe Ullger, the head of Gibraltar's special branch, and a number of other British intelligence officers. The M15 surveillance team was on the streets of the town along with his own special branch men, detective constable Charlie Huart was sitting with the Spanish police at the border control post waiting for the bombers and anyway Canepa had been told that the car bomb almost certainly would not be brought across the border until Monday at the earliest. So, at 12.30 p.m. on Sunday Joseph Canepa went home for lunch.

The police commissioner felt reassured. Only twelve hours ago he had taken part in a midnight briefing for everyone involved in Operation Flavius. The name was appropriate. Gaius Flavius had been a fourth-century Roman politician who had published the rules of legal procedure for the Roman empire and this operation, Canepa was determined, was going to be run according to proper rules. He had already laid down the objective: to arrest the offenders, disarm them and defuse the bomb. True, he had included the rider that this might involve 'the use of lethal force for the preservation of life'. That, however, was seen as an outside possibility. Canepa had got the impression that there was almost nothing that the intelligence people didn't know. And where gaps had had to be filled the conclusions were based on a blend of fact and careful assessment. He had laid out the position clearly as he saw it:

- There was to be an IRA attack on the band and guard of the Royal Anglian Regiment in the assembly area in the centre of town on the morning of Tuesday 8 March.
- The terrorists would use a large car bomb designed to kill as many soldiers as possible.
- The car bomb would be a button job triggered by a sophisticated RCIED (remote-control improvised explosive device) from some distance away. The Provisional IRA began using radio control to detonate bombs in 1972. Initially, they used simple control equipment normally used to guide model aircraft but they soon realized that random signals could also detonate their devices. The army had developed a radio that could rapidly scan the frequencies and emit a pulse that would blow up IRA

157

bombs. This accounted for the deaths of a number of IRA terrorists, who officially died while preparing a bomb in in what were described as 'own goals'. The IRA then designed a simple but reliable encode/decode system which remains in service. They have used a wide variety of transmitters and receivers but have now standardized on a VHF system operating in the two-metre band. They have used remote control in more than 200 attacks, including Warrenpoint in 1979 where eighteen soldiers were killed, the murder of Lord Mountbatten in 1979, the bombing of the Household Cavalry in Hyde Park in 1982 and the murder of Lord Justice and Lady Gibson in 1987. A timer would not be used. Such a device could not be relied upon to go off at precisely the right moment and an ill-timed explosion would kill civilian bystanders thereby diminishing any propaganda coup.

● The active service unit would not use a blocking car to 'reserve' a parking space in readiness for the car bomb. The reasoning here was that a blocking car would mean making two trips across the border and would need two drivers – an unnecessarily risky procedure. When the car came across the border it would be the one carrying the bomb.
● The active service unit was ruthless, would be armed and, if compromised, would not hesitate to use force.
● There was no surveillance on the Spanish side of the border and while the bomber was expected to come on either Monday or Tuesday, everyone should be on the alert from that morning forward.

In theory it was a masterly exposition of the dangers facing the relatively small group of men preparing to meet one of the IRA's most experienced active service units. Unfortunately it was wrong on three vital points. The three terrorists were not armed, the car driven into

Gibraltar on that Sunday was a blocking car and it contained no explosives, and finally the bombers intended using a time bomb and not a radio-control device. Had the SAS known the three were unarmed, that there was no bomb and no radio-control device, all three would be alive today. However, in dealing with terrorists the first rule is always to assume the worst and that their operations will follow logically from things learned in previous operations. The assessment in Gibraltar was wrong but it was also logical, and on the evidence they had available none of the people involved in Operation Flavius could have drawn any other conclusion than the one they did.

The advisory group had now to decide what to do when the bombers appeared. There were three options as set out by the commander of the SAS in the field, soldier E. The first was to arrest the bomber or bombers as they crossed the border. This was easier said than done. The Spanish police had lost all three bombers at Malaga airport and there was little evidence that they would pick them up again before they reached Gibraltar. The police had no idea at this stage what kind of vehicle the terrorists were travelling in. This meant that there would be no warning when they reached the frontier and the guards would have around ten to fifteen seconds to make a positive identification and arrest. Not an easy thing to do, especially when the suspects are sitting in a car, probably armed with guns and a button-device bomb. Equally it would be impossible to slow the frontier control down to give more time for surveillance without arousing suspicion.

The second option was to arrest the bombers as they emerged from the car in the main square, but once again this depended on advance warning and positive identification. In the end the option adopted was the third: to arrest the terrorists once they had been positively identified and were away from the car bomb. To this end the two groups of two SAS soldiers were briefed for a second time in the early hours of Sunday. They were reminded of the characteristics of the ASU, that they were ruthless and usually armed and if confronted would use whatever weapons they had. They were also likely to use any button jobs they had on them.

Having outlined the enormity of the task facing his men, Soldier E then went on to set the limits to their powers in dealing with that danger.

The rules of SAS engagement have always been a well-kept military secret and the ones used in Gibraltar were, it emerged, little different from those used in Britain and Northern Ireland. There were four in all. Control must remain with the police unless passed to the military, and that demanded a hand-over certificate from the senior policeman to the senior soldier, signed by both. The aim of the operation was to arrest Farrell, McCann and Savage. Minimum force was to be used at all times. And finally, as it was explained at the inquest in September, 'firearms could be used but only if those using them had reasonable grounds for believing that an act was being committed or about to be committed which would endanger life or lives and if there was no other way of preventing that other than with firearms'. The rider to this was most important. 'A warning could be used if firearms were to be used, but if a warning would increase the threat, then the warning could be dispensed with'.

The SAS men were about to face three opponents whom they knew to be ruthless, whom they believed to be armed and had the means to blow up a large part of the centre of Gibraltar, whom they suspected would do just that if challenged, and against whom they could use firearms if they thought their lives or those of other people were in danger. In these circumstances there was almost an inevitability about the events that followed.

On the afternoon of Sunday 6 March the MI5 observer is enjoying a quiet cigarette in the main square when he comes face to face with a man in a pin-striped jacket and jeans whose photograph he is certain he has been looking at all week. It is 2.30 p.m. Sean Savage has arrived on the Rock. With a start the watcher turns, walks away and then on the surveillance radio net reports back to the joint operations room the possible sighting of Savage. At almost the same moment other watchers stationed at the border report that two persons answering the descriptions of Farrell and McCann have entered Gibraltar on foot and are making their way past the airport terminal towards the town centre. Meanwhile Savage has wandered back across the square to a Renault 5 parked in slot 2 with its nose to the wall and its boot facing into the area where the band halts at the end of its parade. He unlocks the car and fiddles with something inside for two or three minutes. No watcher is near enough to see what he is doing but the significance is obvious. The waiting is over.

The two SAS teams are now ready and awaiting instructions. Each soldier has brought with him from Hereford a 9mm Browning semi-automatic pistol, 9mm ammunition, and empty magazines. Those magazines are now loaded, each with twelve rounds, and one magazine is in each pistol. The pistol, safety catch on, is down the back of each soldier's trousers, its butt hidden by his jacket. Each soldier carries a radio. One of each pair is tuned into the military command net, the other to the surveillance net. The microphones are held in the collars of lightweight jackets or sports shirts, the on-off switches are taped to their wrists so that transmission involves using only one hand. With each two-man team is an armed Gibraltar policeman. His job is to act as a witness and to undertake the formal arrest of the IRA bombers.

Soldier E opens up the tactical radio net and talks to

his two teams. A and B are told to take up positions near the convent to watch over Savage, C and D to make for the Queen's Hotel to pick up the suspected Farrell and McCann. At 2.30 p.m. Charles Colombo, the deputy police commissioner, telephones his boss at home and lets him know that 'persons matching the description of Farrell and McCann' have crossed into Gibraltar. The minutes tick by, the watchers watch, Savage hangs around the square, the car sits menacingly in the corner, Farrell and McCann are still heading steadily on foot into the centre of town. But the identification is not positive; Canepa decides to stay at home for the time being.

Suddenly the active service unit is in the square not a stone's throw from the operations room. It is 2.50 p.m. Farrell and McCann have joined Savage and all three walk along the line of parked cars. They stop and look at the Renault. Not for long, but long enough to chill the MI5 watchers. Now there can be no doubt. Identification is positive and the time to strike has arrived. At 2.55 p.m. Colombo agrees to the arrest and, verbally for the moment, passes the civil power to Soldier F and the military. At that moment the three bombers turn away and head out of the square past soldiers C and D. The soldiers report in with a positive identification and set off in pursuit. Soldier E in the operations room tells them to hold back. Control is once more back with the police.

Once again Colombo telephones the commissioner. He tells him there is now little doubt that the bombers are in town. Canepa returns to the operations room. It is 3.00 p.m. as he comes through the door. Colombo tells him that the three suspects are heading south towards the Alameda Gardens, that they now believe Savage parked the car in the square as early as 12.50 p.m., and that an arrest had almost been made when the three terrorists met up at 2.50 p.m. Canepa decides to hand control to the military again, then, just as suddenly, changes his mind. He tells the surveillance team: 'I want one final check on identification.'

A fan stirs the air in the crowded operations room; suddenly the message comes through that the suspects are heading back from the Alameda Gardens. At 3.25 p.m. they come back through the Southport Gate into the assembly area. As though irresistibly drawn towards it, they go once again and examine the car. Then they're off – heading north along Main Street and towards the border and moving fast. All doubts have vanished, these are the bombers. On the commissioner's instructions Soldier G, the bomb disposal expert, hurries out of the operations room and across the square to the Renault. He has two minutes to decide whether or not it contains a bomb.

Savage fiddled under the dashboard when the car was parked, didn't he? That suggested a bomb. At the briefing G has been told that a blocking car was most unlikely. 'When the car came it would be for real.' True, it is at least twenty-four hours earlier than expected, but why doubt the intelligence people who have been absolutely right so far? On the other hand there is no sign of the car being uneven on its springs from the weight of a bomb. But nowadays the IRA are expert in distributing the explosive evenly so as to conceal a tell-tale tilt. And then there is the aerial. For G that settles the matter. The car is relatively new, the radio aerial is old and rusty, suggesting it might have been fitted at a different time and could be used to detonate a bomb. Clearly this is the business end of the 'button job' he had been briefed about. G reports back to the commissioner. He has not been able to examine the car carefully, but in his opinion the car could contain a bomb which, if detonated, would certainly kill or maim most of the people currently in the square. At 3.40 p.m. exactly the commissioner of police signs the document handing the power of arrest to the military.

As Soldier G is reporting back to the commissioner,

Soldiers C and D, with their arresting officer and a full surveillance team, are following the three terrorists along Line Wall Road. A and B are making their way to Smith Dorrien Avenue ready to join C and D and the surveillance team as the terrorists pass on their way to the border. C and D report back to control, they are confident of carrying out the arrest. Soldier E passes control of the operation to the two teams on the ground who continually listen on both the surveillance and command radio nets. As this happens soldiers A and B emerge from the Landport tunnel; and there ahead of them standing and talking at the junction of Smith Dorrien Avenue and Winston Churchill Avenue are the three members of the IRA active service unit.

Up the road at the border Detective Constable Charles Huart of the special branch is still sitting with photographs of Savage, McCann and Farrell ready to spot them should they decide to cross into Gibraltar. He can't actually see the border: the Spanish police have put him in a room with a closed circuit television screen on which they project the passports of people crossing. After seven hours Huart is beginning to suspect that the Spaniards are being less than scrupulous in showing him all the passports and he is getting irritable. At around 3.30 p.m. someone remembers him. He is told the terrorists have been sighted and he should come back to Special Branch headquarters. He gets on his motorbike and heads for town. Suddenly there on the corner are all three terrorists, whom he recognizes instantly. At that moment the three separate. Farrell and McCann continue to head north towards the border. Savage turns round and heads back the way he has just come. Huart swings his motorbike into the nearby housing estate, stops and watches as the final drama unfolds.

Inspector Louis Revagliatte is, in his own words, in charge of routine policing on the Rock. Operation

Flavius is most definitely not routine and no one has told Revagliatte what is going on. Just before 4.00 p.m. he is sitting in a police car with three other officers in a line of traffic waiting to turn from Smith Dorrien Avenue into Winston Churchill. Suddenly, he gets a call to return to police headquarters urgently. He puts on the car siren, swings across the road and drives up the left-hand side of Smith Dorrien. (This turned out to be one of the great ironies of the case. The message calling him back to HQ was so that there would be a car available to take the terrorists to jail after they had been arrested. In fact, by turning on his siren, the policeman may well have caused the deaths of the terrorists.)

Soldiers A and B say they did not hear the siren, but when you are approaching IRA bombers it is possible that all you ever hear is your own heart beat. Huart heard the siren and the gunshots 'more or less at the same time'. Officer P, one of the armed Gibraltar policemen, told the inquest that both terrorists became hyperactive when the police car siren 'blared behind them'. Stephen Bullock, a local barrister who was out walking with his family that afternoon, was certain he heard a police siren followed in almost a split second by the first burst of gunfire. An MI5 surveillance officer told the inquest that she heard a police siren seconds before Farrell and McCann were shot.

Siren or not, something makes McCann turn and look behind just as soldiers A and B are closing in on him and Farrell. In the words of Soldier A: 'As I was getting close to Farrell and McCann, McCann looked over his left shoulder. He looked over but his body did not turn. He was smiling at the time. I was behind him, approximately ten metres . . . he looked straight at me. We had eye to eye contact. We looked directly at each other. The smile went off McCann's face . . . it is hard to describe but it is almost like McCann had realized who I actually was . . . I was just about to shout a warning to stop and at the same time I was drawing my pistol. I went to say "Stop" . . . I don't know if it actually came out, I honestly don't know. He looked at me and all of a sudden his right arm, his right elbow, moved, in what I deemed an aggressive action, across the front of his body. I thought mad McCann was definitely going for the button. To me the whole worry was the bomb in the assembly area.'

Soldier A, knowing that McCann is considered to be the IRA's greatest exponent of close quarter combat, now has his Browning in his hand and fires one round into McCann's back. He sees Farrell out of the corner of his eye 'going for her bag'. By now the two soldiers are just three metres away from the bombers. Soldier A assumes Farrell is going for the button. He shoots her in the back and then shoots McCann three more times, one in the back, two in the head. In all, A fires five times.

Soldier B is concentrating on Farrell. He sees McCann turn and look back but not clearly. Suddenly A is firing. He draws his own gun as Farrell draws a large shoulder bag across her body. 'In my mind she made all the actions to carry out the detonation of a radio-controlled device. Uppermost in my mind was the lives of the general public in Gibraltar, my comrades, and the public in the area . . . I fired one or two shots at Farrell and then switched fire to McCann . . . in all seven shots.'

Those twelve shots ring out as Savage is heading for the Landport tunnel with soldiers C and D following him closely. Savage spins round very fast and as he does so C shouts 'stop'. He also sees Savage put his right hand down to the area of his jacket pocket. C already has his gun in his hand. 'I fired because I had been told there was a bomb in the Ince Hall area and it would be detonated by one of the three. It was most likely to be Savage because he had earlier been seen in the vehicle playing around with something, and it had been made clear in my mind that it was definitely a bomb because there was an old aerial on a relatively new car.' C and D carry on firing until Savage has gone down and is 'no longer a threat to initiate that device'. Soldier C fires four rounds to the chest and two to the head. Soldier D fires nine rounds. In all, Savage is hit no less than fifteen times.

At 4.06 p.m. soldier F hands over control of the military operation on the colony back to the police commissioner. The military operation has been running for just twenty-six minutes. Three of the IRA's most ruthless operatives are dead. Could anyone complain at that?

CHAPTER 11

Fall-out

News of the killings hit Britain's newsdesks at about 5.30 GMT in the afternoon of Sunday 6 March 1988. The newsmen whose job it is constantly to scan the wire services saw the terse message from the Press Association on their computer screens: 'Three IRA terrorists shot dead in Gibraltar.' Political and defence specialists were alerted and started ringing round their contacts. The Northern Ireland correspondents were briefed and set to work. Correspondents in Gibraltar and Southern Spain were roused from Sunday siestas. Whitehall and the Ministry of Defence began answering questions as best they could, and while the news stories were assembled for Monday's morning papers, still more reporters were booked on that evening's planes to Gibraltar, if possible, or if not to Malaga some 130 kilometres up the coast of the Costa del Sol. It was, everyone agreed, a remarkable story and clearly a famous victory for the security forces.

Monday morning's newspaper stories set the tone: 'IRA bomb gang killed in Gibraltar – parade carnage plot foiled as 500-pound device in car defused'. Later in the morning the IRA admitted that the three dead terrorists were an active service unit on a bombing mission. The announcement was made by Danny Morrison, the hardliner who favours more military action against the British. Relations

168

between him and Adams have been strained recently and it is understood the announcement was made without consulting Adams. By prematurely confirming the presence and plans of his men, Morrison played into the hands of the British. In the inquisition after Gibraltar, this one incident caused considerable tension between Morrison and Adams and further soured the relationship between the two. No one had ever doubted the IRA's involvement, but confirmation made the victory that much sweeter. It seemed an open-and-shut case of justifiable homicide. What no one could have predicted was that the killings in Gibraltar would trigger a whole new campaign in the war between the SAS and the IRA. That within a matter of weeks a deranged loyalist gunman would murder three people and injure sixty-eight more as the three Gibraltar terrorists were being buried at Belfast's Milltown cemetery; that two British army corporals would be brutally murdered for wandering too close to the funeral of one of the republican victims; that the British government would be in raucous dispute with both the IBA and the BBC; and that the SAS would, in effect, be put in the dock, not as heroes, but murderers.

That Monday afternoon, 7 March, almost to the minute twenty-four hours after the shooting, Sir Geoffrey Howe, the British Foreign Secretary, rose in the House of Commons to tell MPs what had happened in Gibraltar. The government's account of the hours before the shooting was clear and unequivocal. Just before 1 p.m. on Sunday, Sean Savage was seen parking a white Renault 5 in an area where, each Tuesday, the military bandsmen gather before and after the changing of the guard. Mairead Farrell and Daniel McCann crossed the border from Spain on foot around one and a half hours later and met up with Savage in the centre of town at around 3 p.m. At about 3.30 p.m. on their way to the border they were challenged by the security forces and shot dead. The MPs listened in silence. Sir Geoffrey continued: 'Their [the terrorists'] presence and actions near the parked Renault gave rise to the strong suspicion that it contained a bomb, which appeared to be corroborated by a rapid technical examination of the car.' The problem was, as Howe then admitted, that the car did not after all contain a bomb, and, as he further admitted, all three terrorists were subsequently found not to have been carrying arms.

Monday morning's newspaper stories were suddenly wide of the mark; it was no longer an open-and-shut case of armed terrorists on a bombing mission, with the bomb in place, being shot dead after a challenge to surrender.

It took some time for the full import of Howe's words to sink in. Indeed George Robertson, Labour's foreign affairs spokesman, in replying to Howe's statement, continued to speak about this 'enormous potential car bomb placed opposite both an old folks' home and a school.' Before this and on behalf of the Labour leadership he had congratulated 'those responsible for what appears to have been a well-planned operation.' There was, however, a harbinger of events to come. Eric Heffer, the left-wing MP from Liverpool, asked Howe 'to explain why these people . . . were shot and killed when it was admitted that they were not carrying guns and had not planted any bombs in Gibraltar?' Howe played a straight bat. He did not know how the security forces could have acted in any other way than they did, given the circumstances.

What he could not escape, however, was the fact that until that moment the security forces had believed a 500-pound bomb had been planted in the car. This information emerged out of the confusion immediately after the shootings. Today, no government department can say where the '500-pound bomb' came from. If the government could be that wrong on such a straightforward matter of fact, was there any reason to trust what else was being said? Wasn't this just 'shoot-to-kill' all over again? The fact that on Tuesday the Spanish police discovered a 141-pound car bomb in an underground car park in Marbella did nothing to quell the disquiet. It was too late, the damage had been done. The hare was running and there was no shortage of people ready to join the chase.

The government knew there might be criticism. What it did not expect was the source of the first salvo or the speed with which it arrived. Under its new editor, Max Hastings, *The Daily Telegraph* had of late abandoned its inclination to follow the Tory party through thick and thin. Now, on Tuesday morning, it ran a hard-hitting leader that asked the government why it had given a succession of contradictory accounts to the world about Sunday's events. 'Unless it wishes Britain's enemies to enjoy a propaganda bonanza, it should explain why it was necessary to shoot dead all three terrorists on the street, rather than apprehend them ... it might be deemed politically prudent to give ill-wishers no further grounds to suggest that the authorities operate a shoot-to-kill policy against terrorists.'

In any society that cherishes its freedoms as much as the British there are bound to be questions asked when a criminal, no matter how appalling his or her past crimes and future intentions appear to be, is shot dead rather than brought to trial. While the arrival of terrorists ready to murder indiscriminately has blurred such matters, public unease remains unless four quite specific conditions are met: those shot dead should be known terrorists, they should be on a murder mission, they should be armed and finally they should either be seen to fire first or be seen to respond aggressively to an order to surrender.

The ambushes at Loughgall and at Omagh met all these conditions and the IRA publicity machine – one of the world's most sophisticated – found it hard to capitalize on the deaths that occurred in those two episodes. There were no martyrs to be paraded before the waverers in Ulster's nationalist camp. Gibraltar was different. Matters were less clear cut. Howe had admitted the three terrorists were not armed and there was no bomb in Gibraltar.

Among the first to voice doubts about the way the SAS had conducted itself was Ireland's prime minister, Charles Haughey. A statement issued after the Tuesday cabinet meeting in Dublin said: 'The Government is gravely perturbed that three unarmed Irish people should have been shot dead in Gibraltar . . . when it appears from reports, that they could have been arrested.' Later in the Irish parliament Haughey re-emphasized his view that the shootings of unarmed civilians 'should be unacceptable to democratic governments everywhere'.

Mrs Thatcher became increasingly furious at the questioning of the Gibraltar operation. In parliament she defended the SAS and explained how relieved she was that an outrage that would have put even the Enniskillen poppy day massacre in the shade, had been prevented. The next day *The Sun* newspaper, with its unerring ability to reflect the government's gut reaction to political events, came back with a tough little leader comment: 'Charley Haughey is gravely perturbed . . . Well, we are perturbed that his citizens are roaming the world trying to blow up our innocent people.'

By Thursday the left wing of the Labour party had come out of the shadows: thirty-five members of the Left put their names to a Commons motion describing the shooting of the three IRA bombers as an 'act of terrorism'. The motion was sponsored by Eric Heffer, in effect repeating his performance of the Monday before. He condemned the actions of the SAS as 'tantamount to capital punishment without trial'. Support came from Tony Benn and Ken Livingstone. Neil Kinnock and his senior colleagues let it be known that they had little sympathy with the left-wing motion, but there was, however, one significant exception. Kevin McNamara, a Protestant and Labour's spokesman on Northern Ireland, was quick to question his party's stand on the issue. He accused the government of playing into the hands of the IRA by creating three 'martyrs'. 'The rule of law and British justice is at stake over these matters and the government is refusing to face up to the enormous damage which has been done in Northern Ireland and to British prestige in the world.'

It was increasingly clear that the government was in danger of forcing itself into a corner. In his statement on the Monday after the shootings Howe had said that an inquest would be held in Gibraltar. So convinced was Mrs Thatcher that the SAS men had done no more than their duty that she had refused all attempts to hold a judicial inquiry, which would have implied that the government and the SAS had been wrong to act as they did. She stuck steadfastly to the view that an inquest was the correct way to handle matters. However the rising winds of criticism made it clear that even this might severely embarrass the authorities. Any verdict other than 'justifiable homicide' would be seen as a victory for the IRA.

Nor were the noises from Gibraltar and in particular from the office of the coroner, Mr Felix Pizzarello, helping matters. He seemed to be very much his own man in insisting that the SAS team turn up and give evidence. This Mrs Thatcher saw as madness. In her eyes the courageous men who had saved the colony from major carnage would be publicly identified and risk death at the hands of the IRA. Moreover, once identified, they would be unable to continue their chosen profession as members of one of Britain's most important regiments – a regiment, moreover, that could be relied upon to carry out its duties with efficiency and secrecy. In other words, as far as the government was concerned, there was no case to answer. Unfortunately, neither Thames TV nor the BBC shared Mrs Thatcher's view of events.

The government's relations with the broadcasting authorities have never been easy. There had been rows over the coverage of the Falklands war and of Ireland. Norman Tebbit launched an attack on the BBC over its reporting of the American bombing raid on Libya. Mrs Thatcher further took the view that the IBA was weak and ineffectual for failing to come to grips with trade union power within commercial television. Now the biggest row of all was about to break, but this time the government played all its cards wrong and gave to the IRA the very thing it had always fought against – the 'oxygen of publicity'.

There are very few greys in Mrs Thatcher's world and when it comes to terrorism, none at all. In an earlier row when the BBC refused to hand over untransmitted film of an IRA funeral, she said, 'Either one is on the side of justice or on the side of terrorists'. This dictum she now applied to the television companies making documentaries on the Gibraltar shootings in the belief that they could be persuaded not to show the films with the inquest pending.

The Ministry of Defence knew that Thames TV were making a film about the shooting – the Ministry had been asked in mid-March for help in assessing the damage likely to be caused by 141 pounds of Semtex. Not surprisingly it refused. In all the TV team spent several weeks in Gibraltar searching for and interviewing new witnesses to the shooting. As the programme neared completion contact was made with the MOD, and the two parties met on the evening of Thursday 21 April. By the following Monday the Prime Minister and her senior colleagues knew that Thames' film *Death on the Rock*, scheduled to appear on the following Thursday in the 'This Week' slot, would suggest that Farrell and McCann were about to surrender when they were shot outside the petrol station. In other words Thames was about to go to the very heart of the government's justification for what happened on that Sunday as expounded by Sir Geoffrey Howe the following day: 'When challenged, they [the terrorists] made movements which led the military personnel . . . to conclude that their own lives and the lives of others were under threat.'

The first Lord Thomson of Monifieth, the chairman of the IBA, knew of any government alarm over the film was when he received a telephone call from Howe at 5 p.m. on Tuesday, two days before scheduled transmission.

Howe had already consulted the lawyers at the Foreign and Commonwealth Office to see if there were legal grounds for getting an injunction against the film. It was out of the question. The inquest was to be held abroad and the rules of contempt did not therefore apply. Hence the call to Thomson – would he postpone the film until after the inquest?

Ever since the row between the government and the BBC over the 'Real Lives' programme, which included an interview with one of the IRA leaders, television executives have handled films about Ulster and the IRA with great care. On this occasion, and well before the Howe telephone request, Thames had asked the IBA if it would like to see a tape of the programme. Normally the preview would be carried out by a member of staff at the IBA headquarters in Knightsbridge. On this occasion Thomson and his two most senior colleagues – John Whitney, the Director General and David Glencross, the television director – watched the film. They could see no reason for acceding to the foreign secretary's request. There was, as Howe had already discovered, no legal obstacle to transmission; and as for influencing the inquest, the television reporters were doing no more than what newspaper journalists had been doing ever since the shootings in March – trying to discover what actually happened. Hadn't the government given its own version of events immediately after the shooting? Why should television be singled out as different?

The fact is that Mrs Thatcher does see a difference between print and television journalism – a difference that demands greater responsibility from television because its impact on the public is so much greater. The IBA executives watched the film on Wednesday and on Thursday morning gave Thames the go-ahead to transmit that evening. Sir Geoffrey informed the Cabinet of the IBA decision at the Thursday morning cabinet meeting. Clearly up to this point it might have been possible to keep the disagreement between the government and the IBA private. However, at 1.45 p.m. Lord Thomson received a telephone call from his office at the Westminster restaurant where he was lunching. The message was far from pleasant. The Foreign and Commonwealth Office were briefing the lobby correspondent about the IBA's refusal to postpone the Gibraltar film until after the inquest. The government insists it did this only because Thames had begun to leak a story suggesting that the government was once again wielding the big stick against television. Where the truth lay is impossible to judge. However the IBA's response was to issue a press statement that could hardly have been better calculated to make Mrs Thatcher's blood boil. A postponement would only have served to give the IRA more publicity. This was a pointed reference to her own remarks after the 'Real Lives' row with the BBC in 1985.

The programme was transmitted on the evening of Thursday 28 April, and the government's worst fears were confirmed. The 'This Week' reporter, Julian Manyon, was adamant on camera: 'We have interviewed four key witnesses to the shootings . . . they say that the British soldiers opened fire without warning and none of them saw the IRA bombers make any threatening movements.' The programme went on to say that the SAS fired further shots into the terrorists' bodies as they lay on the ground. This was strong stuff and the viewer was left in no doubt that the killings were cold-blooded and unnecessary and, if true, the soldiers would have to be charged with murder. America's prestigious *Time* magazine was certainly convinced by the programme's evidence. It told its readers the following Monday that 'the film shows four witnesses who say that two of the suspects . . . had raised their hands in surrender when commandos in plain clothes shot them.' Dublin was outraged. After the broadcast it called on Britain to bring the SAS to justice.

The next morning's newspapers left no doubt how the government felt. It was 'trial by television'. Sir Geoffrey Howe went on the record: 'It is a matter of simple common sense that the televising of interviews with one or more potential witnesses before an inquest takes place must carry a serious risk of prejudicing the proceedings. To disregard that risk is irresponsible.' Thames TV was accused by one senior government spokesman of acting as judge, jury and prosecuting counsel.

Mrs Thatcher shared all these views and more. In an interview with Japanese TV on Friday 29 April she said that 'trial by TV or guilt by accusation is the day that freedom dies'. She went on to say that her feelings about the programme were 'much deeper than being furious.' Lord Thomson was equally angry. 'Trial by television,' he told *The Observer* newspaper, 'is a fallacious phrase. It is not trial by television at all. It is no different from much of print journalism investigation which quite rightly goes on.' He added, somewhat injudiciously in view of later events, 'I think the events of this week will enhance our reputation with politicians of all persuasions.'

Just one week later the whole issue came roaring back into the public domain when the BBC announced that it too intended broadcasting a documentary on the Gibraltar killings in its 'Spotlight' programme. This is transmitted in Northern Ireland only, and while the audience was more limited than for *Death on the Rock*, which had been net-worked to all the commercial channels, nonetheless Sir Geoffrey Howe once again deemed it a threat to the impartiality of an inquest. He telephoned the new BBC chairman, Sir Marmaduke Hussey, on the Wednesday with a request that the programme should not be transmitted. The BBC, still smarting from the 'Real Lives' row, took its time reaching a decision. Just seventy-five minutes before the programme was due to go on air and twenty-four hours after he had made the request, Sir Geoffrey received his answer. As with the IBA it was 'no'.

Earlier that day the prime minister had once again made her feelings abundantly clear. She told MPs that in the past a high sense of responsibility meant that Britain has 'escaped the horror of trial by press, television and radio. It is that sense of responsibility which does not seem to be there now.' She went on to accuse the BBC of 'flouting the rule of law'. Roy Hattersley, Labour's shadow Home Secretary, riposted with a statement congratulating the BBC for withstanding 'the government's authoritarian attempts to censor television'. The IRA leadership must have loved it as the battle between the government and the broadcasters kept the terrorists' names and their cause before the public day after day.

The press was divided over whether or not the programmes should have been shown. The tabloids fell into their predictable political pigeon-holes with most of them sharing the government's outrage, but the quality papers were more cautious. Once again *The Daily Telegraph* refused the government succour. In its leader comment following the transmission of *Death on the Rock* it admitted that the programme was politically most inconvenient. 'Yet it is difficult to see how this justified the government's request to the IBA. Thames Television's programme and the evidence of its witnesses may well prove to have been entirely without merit. But the film-makers have broken no law'. In other words there was no good reason why Thames should not make such a programme and show it, but equally there was an obligation on Thames to make sure the programme was accurate. Only one newspaper tested that obligation with any conviction – *The Sunday Times*.

The Sunday Times Insight team had had dealings with Julian Manyon and Thames before. In 1981 Insight had investigated a programme made by Manyon for Thames' 'TV Eye' programme about Czech dissidents. The investigations revealed distortions – a charge that was later upheld by the Broadcast Complaints Commission, which directed Thames to broadcast a summary of the complaints and the Commission's findings. Thames initially issued writs against *The Sunday Times* but pulled out shortly before the date of the trial, agreeing to pay £60,000 towards *The Sunday Times'* costs, to apologize on television and to issue a statement saying it did not knowingly broadcast a programme that contained lies.

On 1 May, *The Sunday Times* published an article, 'Inadmissable Evidence', criticizing Thames for 'sloppy journalism'. As in 1981, Thames was outraged and, although no writs were issued, Thames' director of programmes, David Elstein, demanded and got the right to reply. In this he said: '*The Sunday Times* claimed that much of our evidence is invalid or self-contradictory, and that our witnesses complain that we have misrepresented them. All these charges are without foundation.'

It turned out in court that a key witness relied upon by the 'This Week' programme was a nineteen-year-old bank clerk, Kenneth Asquez. He had claimed to have seen the shooting of Sean Savage, and Manyon had relied on this statement for the evidence that the SAS men had pumped bullets into Savage as he lay on the ground. In fact, four months later, Manyon sat in court to hear his star witness admit under oath that he had not seen the shooting. 'I made this up,' he told the inquest.

Trial by television had suddenly become trial of television. *The Times* of 24 September remarked: 'Journalism which advocates further penetrating inquiries by official bodies [Thames had demanded a judicial inquiry rather than a simple inquest] needs to apply its own prescription to itself.'

Asquez's revelations brought audible sighs of relief from Whitehall. By admitting he had lied he effectively destroyed the credibility of the whole programme. He also put televison on trial and made it look as though the government's attempts to stop the television programmes were somehow justified.

Asquez had not appeared on the programme. Manyon explained to camera that he wished to remain anonymous. But he had made a 'detailed statement' to a local lawyer, Christopher Finch, who represented the television company. The core of the statement was just seventy-two words long: 'The man on the ground was lying on his back, the man standing over this man had his foot on the man's chest. I could see that he also had a gun in his hand. I then saw the gunman point his gun deliberately at the man lying on the floor and fire two or three times into him at point blank range. I was horrified by what I saw.' And so were the viewers by what they heard; this was cold-blooded savagery.

In fact the statement was worthless. Asquez had originally made a short unsigned handwritten statement, which he had given to Major Bob Randall, a local man who had approached him on behalf of the TV programme. He was then seen by Finch in his office when he said that he wished he had never made the statement and had no wish to make another. After Asquez left his office, Finch drafted out his recollection of the conversation in the form of a second statement, which he then passed on to Thames. They later approached Asquez who refused to sign it. Even so, the programme used it as its main source.

During the inquest Asquez admitted that he had not actually seen the shootings. It was over this issue that the government recognized that they made a serious tactical mistake during the inquest. The government believed that Asquez's acknowledgement from the witness box that he had seen nothing of the shootings should have been sufficient to undermine his previous statement completely. However, by the time the inquest had ended, the lawyer for the families of the dead terrorists, Mr Paddy McGrory, had sufficiently muddied the waters so that even the coroner appeared unsure of what Asquez did or did not see. The apparently unresolved questions centred around Asquez's report of seeing the SAS men wearing berets, and identifying themselves after the shootings. In fact, Asquez had not seen this but heard about it from gossip in the town and from the media. The government had other people available to testify who were in the car with Asquez. If called, they would have said that they saw nothing, and it is this failure by the government lawyers to call them to corroborate Asquez's statement from the box that was the mistake.

As the evidence ended it was left to Felix Pizzarello, the Gibraltar coroner, to pose the crucial question: 'Why did Thames TV put it [Asquez's statement] out as fact when it was just a draft?'

Day fourteen should have been the high point of the inquest for the producers of *Death on the Rock*. The day their 'witnesses' went into court. Asquez's evidence had already turned it into a nightmare, but the ordeal was not yet over. The second of Thames' star witnesses was Carmen Proetta. Following her appearance on the film, Proetta had been mercilessly hounded by Britain's popular press. A whole series of unproven but unpleasant allegations were made by the media about her personal and professional life. Almost inevitably, therefore, there was a good deal of sympathy for Proetta and a willingness to believe what she said. Unfortunately for Thames, when she went into the witness box under heavy cross-examination she was less certain about what she had seen and heard than when she appeared on television.

In the film Proetta had said that Farrell and McCann had put their hands up in an 'unresisting gesture'. Asked if this meant as if they were 'trying to surrender', she answered unequivocally 'correct'. Now she was not so sure. She now replied: 'for me the signal of hands up could mean surrendering or shock'. On television Proetta had been certain that the shots which had killed the two terrorists had been fired by the men who had jumped over the central reservation in the road outside the service station in Winston Churchill Avenue. Now she hesitated and then told the court: 'I didn't notice where the shots came from. I have no idea where they came from. I didn't see any trace of smoke and firing.' More was to come.

Perhaps the most damning charge against the SAS was one of premeditated savagery. This was why Asquez's testimony had been so damaging. Proetta had told the television cameras that she had seen much the same thing happen to Farrell and McCann. She said she saw one of the men bend down and fire at least twice into the heads of the terrorists who were, by then, lying on the ground. Yet when asked in court if she might not have been mistaken, that the shots she heard were not from the men bending over Farrell and McCann, but the sound of the shots that killed Savage, she said simply, 'I don't know'.

At the end of the day John Laws, the barrister representing the Crown, summed up the situation. 'Thames TV were prepared to put to the public as evidence something they only had in unsigned form, from someone they knew had been unprepared to put his name to it.'

At 7.15 p.m. Gibraltar time on Friday 30 September 1988, the eleven-man jury filed back into the tiny courtroom and announced their verdict. By a majority of nine to two the verdict was lawful killing. The government had got the verdict it wanted and the SAS could now return to the shadows, and to the job of defeating the IRA.

CHAPTER 12

The Way Ahead

The battles to come between the IRA and the SAS and security forces will be ferocious and unrelenting. No one is in any doubt that there are greater atrocities and more bloody encounters on the horizon. What has gone before may pale into insignificance, for despite setbacks the IRA has become more proficient, more deadly and certainly more dangerously armed.

British intelligence's initial joy over the French seizure of the *Eksund* in 1987 was shortlived. Investigations since the capture of probably the IRA's biggest and most potent arms shipment by French customs off the coast of Brittany have revealed the frightening extent of the terrorists' armoury.

For before the *Eksund* was stopped, four previous shipments of arms from Colonel Gadaffi had got through to the Provos. Those four smaller shipments together equal the 150 tonnes of arms found crated aboard the *Eksund*. Malta was the base used by the IRA, apparently plying an innocent trade between the Mediterranean and northern Europe. Each voyage, however, involved a short rendezvous at sea with a vessel out of Tripoli, where Gadaffi's men, under the supervision of a Libyan intelligence officer called Azur Nasser, quickly loaded guns and explosives aboard. The four previous shipments were dropped off at a small beach near Arklow in County Wicklow, fifty miles south of Dublin.

The first was believed to be aboard the *Casmara*, later renamed the *Kula*, a small converted fishing boat purchased with £45,000 from the IRA's burgeoning bank accounts. It sailed in August 1985 and loaded five miles out of Valetta harbour with seven tonnes of arms; among the cargo were seventy AK47 assault rifles and a consignment of Taurus pistols. The *Casmara*'s second voyage was in October that year. Now called the *Kula*, it was loaded off Malta, this time with ten tonnes of arms including 100 AK47s, ten machine guns and seventy boxes of ammunition. But it was the *Kula*'s third sortie in July 1986 that the IRA had been waiting for.

The shipments were initially deliberately small in case interception by government agencies along the route consigned Gadaffi's gift to the exhibits room of French, British or Irish police stations. The IRA had learnt the lesson of the *Claudia* well; in the mid-seventies British intelligence, sensing an IRA arms shipment was about to begin, set up a sting in which its own operatives provided the ship and the wherewithal for moving the arms, intercepting it *en route* and bagging the IRA crew and the arms they had purchased.

The *Kula*'s third trip reflected the IRA's growing confidence – and complacency. At a different rendezvous a few miles off Malta's twin island of Gozo they picked up fourteen tonnes of arms, including the surface to air missiles that the security forces had always feared the IRA might start to use one day against military helicopters or civilian aircraft. There were four Eastern block Sam-7 launchers and missiles on the illegal manifest. They were landed at a deserted beach, Clogger Strand, a mile south of Arklow, by rubber dinghy. One dinghy was later found abandoned, slashed and sunk in the bay.

That successful trip completed, the IRA purchased a bigger boat, the *Villa*. It was twice the size of the *Kula* and in October 1986, fifteen miles off the Libyan coast in international waters 107 tonnes were loaded on board, including yet more Sam-7 missiles. The boat again anchored at Clogger Strand and during the night around twenty men loaded the arms from the ship on to a fleet of waiting trucks for ferrying inland. It was an impressive piece of organization.

The *Eksund* was to carry the biggest shipment to date. On 27 October a French customs plane spotted the 237-tonne, 130-foot Panamanian registered boat, north-west of Spain heading into the Bay of Biscay. On board were twenty Sam-7s, 1,000 AK47s, 600 grenades, ten heavy duty machine-guns, anti-tank guns, Beretta machine guns, fifty tons of ammunition and two tons of the powerful plastic explosive, Semtex H.

The French spotter plane had been looking for drugs runners and believed the *Eksund* was worth tailing. A French high-speed customs launch shadowed it north to a point just off Roscoff in Brittany where a smaller vessel was seen to rendezvous with it. That was the IRA's mistake. French officials believed a drugs drop-off was in the offing and moved in at 6.30 p.m.

The French had named it Operation Cracker. Maujis Pascal, second in command of French customs cutter Number 36, based at Lezardrieux, led the boarding party. He and two armed officers leapt aboard the *Eksund*, taking the five Irishmen on board completely by surprise. He demanded to see the ship's log book and cargo manifest. 'When we came on to the bridge, we found five people, three had diving suits and one was in battle-dress.' They gave their names as Adrian Hopkins, forty-nine, of Delganey, County Wicklow; Henry Cairns, forty-four, a 'bookseller' from Bray in the same county; William Finn, forty-three, of County Mayo, Dennis Boyle, forty-two, and Edward Friel, thirty-four, both of County Donegal. They claimed there was no log book or manifest and the ship was escorted into Roscoff.

A superficial search uncovered a terrifying booby trap that made the *Eksund* a floating time bomb. At 9 p.m. that night the hold had been opened and twelve bundles of wired explosive were found connected to a detonator. Buried beneath life-jackets on the bridge were five loaded Kalashnikovs and an Erstal machine gun. Pascal had reason to be thankful that his boarding party had surprised the crew, for they were clearly prepared to fight it out and blow up the ship if necessary.

A routine check with Dublin soon revealed the names and passports of the crew were false. It transpired that the man Finn was in fact Gabriel Cleary, described by some sources as the IRA's chief engineer and logistical supply officer. He had served a prison sentence in Portlaoise jail for running a bomb factory. Cleary was an experienced sailor who owned his own eighteen-foot boat and lived in Dublin. By complete chance the British had been able to destroy the biggest arms-smuggling operation in the IRA's history.

As Britain's relations with Libya deteriorated in the early 1980s so Gaddaffi repeated contacts with the IRA that had been dormant for some years. He decided to re-arm the IRA and 1985 and after the American bombing of Tripoli in 1986, his desire for revenge was reflected in the increased size of his gifts. Libya had maintained links with the IRA dating back to the early seventies. The seized boat, *Claudia*, with Joe Cahill, the IRA's second in command on board, was the first proof of Gadaffi's sponsorship. But in 1976 Gadaffi appeared to have cooled towards the IRA, announcing that: 'Our relations with London and Dublin are improving rapidly, the IRA chapter is behind us.' However by 1980 Libyan radio again began to refer to giving 'aid to Ireland'. In 1981, at the time of the hunger strikes, Gadaffi sent a letter to the UN general secretary criticizing the 'British non-humanitarian code of conduct'. In 1982 after Libya had supported Argentina in the conflict over the Falklands, the government-controlled

Libyan media hinted darkly at a more committed attitude towards the IRA in its war against British 'colonialism'. Once more the IRA were, in Gadaffi's eyes, 'freedom fighters struggling against a British military presence'. Shortly after Gadaffi was openly courting the NUM leadership, offering funds for the miners in their strike against government cuts in the industry.

Curiously, the weapons that Gadaffi has supplied appear to be a 'job lot'. For example, much of the ammunition does not match with typical IRA weapons or the guns supplied by Libya and, aside from the Semtex and the ground-to-air missiles, the weapons would hardly head the list of IRA priorities. 'It's as if the IRA man met with the Libyans to ask for weapons and the Libyan contact pointed to a nearby pile of boxes and said "Take that" without having any clear idea what was inside,' said one member of the security forces.

After the *Eksund* was seized, the gun-running operation was suspended as the Irish army and police mounted a huge exercise to find the shipments that had got through. At 7.30 a.m. on 23 November 1987 7,000 soldiers and Garda officers began combing out from the border areas of Ireland and a confidential telephone line was set up in case any had seen something suspicious that might lead the searchers to the arms caches.

In the following days 50,000 houses were searched, 7,573 of them under search warrants. 164 cruisers on the river Shannon were searched along with 775 mobile homes and holiday caravans. More than 30 people were arrested, and the security forces seized 22 rifles, 15 revolvers, 13 shotguns, 4,312 cartridges, 2,277 bullets, 43 detonators, 2 timing devices, one hundredweight of suspected explosive mix, 35 mortar bombs. They located 4 bunkers dug to house arms caches, and 16 'dug-outs'. The biggest was on a farm near Arklow in the south.

A new silage pit had been dug at the farm with the aid of a government grant but police believed the finished result was unsuitable for agricultural purposes. The Garda believed this was the first hiding place of the illegal arms shipments before they were split up and carried north to new hideaways. The pit was in fact empty – the IRA had been tipped off about the impending nationwide hunt.

It was not until Wednesday, 27 January, more than two months after the search began, that the Irish security forces scored their first big find; under the sand at Five Fingers Strand, on the coast at Martin Head in County Donegal, due west of Belfast, they found oil drums packed with weapons and explosives. Routine detective work had registered the presence of a furniture van in the area which belonged to a known IRA supporter. The weapons were brand new, their serial numbers filed down.

Another bunker found later demonstrated the sophistication of the IRA gun-running operation. On the border the Garda found a bunker that contained electric lighting and wooden shelving. A seventy-foot tunnel connected it to another chamber where explosives were believed to have been hidden. The tunnel itself was used as a firing range to test the weapons and adjust the sights and simply for target practice to sharpen up sniping skills.

To date, very little of Gadaffi's gift has been found by security forces in either Ulster or Ireland. What is certain however is that up to three tonnes of Semtex high explosive is hidden somewhere, waiting to be used. Just 140 pounds of it would have devasted Gibraltar's main town. Unless the security forces can find it, the IRA has the potential to build thousands of bombs like that which shattered Enniskillen. The heavy-duty machine guns supplied by Tripoli have been used already, bringing down a British army helicopter in 1988; but the surface-to-air missiles have yet to appear in the gunmen's catalogue of atrocity.

With money pouring in from its increasingly legalized businesses, with a sophisticated new inventory of arms and explosives, and experienced terrorists earning remission on their sentences to rejoin the ranks of active service men, the security forces believe the IRA is as strong now as it has ever been.

A simple calculation clearly demonstrates that the IRA, despite recent setbacks, will be around for a long time to come. While the actual figure of people who will pull the trigger or press a detonator may be under a hundred, there are between 500 and 1,000 activists who are prepared to go on a terrorist operation; there are 1,000 terrorists in jail, some of whom are released every week and who are an experienced manpower pool for the IRA to draw on; the families of those in jail or currently serving form a hard core of around 10,000 people; to which can be added the 87,000 votes Sinn Fein can count on in the north and the 35,000 in the south, out of whom there are perhaps 50,000 people who sympathize with the terrorist activities of the IRA. That bedrock of support is unlikely to change dramatically in the near term. Although the IRA is never going to achieve its aims, terrorism will be a problem confronting British people for many years.

The security forces are bracing themselves for a new campaign of unrivalled ferocity. The security forces believe that aside from their acts of terrorism in Northern Ireland, the IRA have a substantial arms cache already in position in England and an active service unit may already be here waiting to strike. Also, the two IRA terrorists arrested in Germany last August after a number of attacks against British forces in Europe were only providing logistics support for an active service unit. That unit is still operational. There is no doubt that the IRA is on the brink of a new and particularly savage campaign in Europe and Northern Ireland.

The SAS will be at the forefront of the battle to blunt that campaign. And, in the words of one senior member of the security forces: 'There is a bloody time ahead.'

Appendix A

The following is a chronological list of incidents in which the SAS in Northern Ireland were believed to have been involved. Where necessary, the authors have corrected the record.

1974

13 April. The official Sinn Fein newspaper, *United Irishman*, named six men they claimed were SAS commanders based in Northern Ireland. None of the men was in the SAS.

22 April. An eighteen-year-old Pakistani worker at an army canteen was killed at Silverbridge, Co. Armagh by the IRA, who claimed he was a member of the SAS. In fact, he was a tea boy.

20 July. The IRA tortured and killed a twenty-one-year-old truck driver, Brian Shaw, in West Belfast. They claimed he was in the SAS and working undercover. He had in fact left the army two months before he was shot.

1975

1 August. Patrick McElhone shot dead by 12 man patrol from the Royal Regiment of Wales, not the SAS.

1976

7 January. Prime Minister Harold Wilson announced the first official deployment of the SAS to Northern Ireland.

12 March. Leading IRA terrorist Sean McKenna abducted from his cottage in the Irish Republic, taken over the border and handed to the RUC.

15 April. Peter Cleary, senior officer in the IRA, arrested while visiting his fiancée in South Armagh. Shot while trying to escape.

1 May. Seamus Ludlow shot dead just inside the Irish Republic. The IRA blamed the SAS but he was in fact killed by the IRA on suspicion of being an informer.

6 May. Eight members of the SAS arrested after crossing into the Irish Republic. Later tried and fined for possessing unlicensed weapons.

14 July. Patrick Mooney and Kevin Burns arrested in South Armagh. Both were later released in a case of mistaken identity.

1977

16 January. Seamus Harvey shot dead at Culderry, South Armagh, after firing a shotgun at an SAS unit.

14 May. Captain Robert Nairac abducted, tortured and killed. He was a member of the 14th Intelligence Unit and not the SAS.

12 December. Colm McNutt shot dead in Londonderry by a plainclothes soldier when he tried to hijack his car. The soldier was from the 14th Int.

1978

26 February. Paul Duffy shot dead at Ardboe in East Tyrone at a barn containing explosives.

16 March. Lance Corporal David Jones shot and killed and another soldier wounded in a gun battle near Maghera in Co. Londonderry in which Francis Hughes was captured.

10 June. Denis Heaney shot dead in Londonderry after attempting to hijack a car with members of the 14th Int inside.

21 June. Three IRA men, Jacki Mealy, Jim Mulnenna and Dennis Brown, and one passer by shot dead during a bomb attack on the Ballysillan Post Office depot in North Belfast.

11 July. John Boyle, aged sixteen, shot dead in a village graveyard in Dunloy, Co. Antrim. He had no terrorist connections and two SAS men were tried and acquitted of his murder.

30 September. A Co. Tyrone wildfowler, James Taylor, who had no known terrorist connections, was shot dead near Cookstown.

24 November. Patrick Duffy, aged fifty, shot dead at his house in Londonderry. An arms cache was found inside.

1980

2 May. Captain Richard Westmacott shot dead by an IRA unit after he approached a house at 369 Antrim Road, Belfast. Four IRA men were later arrested.

1983

February. INLA member Eugene McMonagle shot dead.

4 December. Colm McGirr and Brian Campbell shot dead as they approached an IRA arms dump in Coalisland, Co. Tyrone. Both men were armed. A third man escaped.

1984

21 February. Sergeant Paul Oram and two IRA men, Declan Martin and Henry Hogan, killed in a gun battle at Carness, Dunloy.

19 October. Fred Jackson killed in the crossfire at an ambush set up to trap an IRA team planning to assassinate a member of the UDR.

2 December. Lance Corporal Alistair Slater shot dead along with John Anthony McBride in a gun battle at Drumrush, Co. Fermanagh.

1985

February. Three IRA members shot dead as they attempted to collect arms from a dump at Strabane, County Tyrone.

1986

26 April. Seamus McElwain shot dead and Kevin Lynch wounded after the two were ambushed while planting a bomb.

1987

8 May. Eight members of the IRA's East Tyrone ASU and one civilian motorist shot dead in an ambush at the Loughgall RUC station in Co. Tyrone.

1988

6 March. Three members of an IRA unit shot dead in Gibraltar.

4 July. Ken Stronge, a civilian taxi driver, shot dead when an IRA unit attacking the North Queen Street RUC Station walked into an ambush.

Appendix B

The rules of engagement for the SAS are the same as those drawn up for ordinary soldiers.

The most detailed example of the rules was submitted for Operation Flavius in Gibraltar. Although some of the names and places would clearly be different, the general guidelines for any SAS operation are identical.

Top Secret

Rules of Engagement for the Military Commander in Operation Flavius.

Objectives

1. These instructions are for your guidance, once your participation in Operation Flavius has been duly authorized. You are to issue orders in compliance with these instructions to the men under your command.

2. You are to operate as directed by the Gibraltar Police Commissioner or by the officers designated by him to control this operation.

Should the latter request military intervention, your objective will be to assist the civil power to arrest members of the IRA, but subject to the overriding requirement to do all in your power to protect the lives and safety of members of the public and of the security forces.

Command and Control

3. You will be responsible to the Governor and Commander-in-Chief, through his Chief of Staff, for the way in which you carry out the military tasks assigned to you. You will act at all times in accordance with the lawful instructions of the senior police officer(s) designated by the Gibraltar Police Commissioner to control this operation.

Ambush

Use of Force

4. You and your men will not use force unless requested to do so by the senior police officer(s) designated by the Police Commissioner, or unless it is necessary to do so in order to protect life. You and your men are not then to use more force than is necessary to protect life; and you are to comply with rule 5.

Opening Fire

5. You and your men may only open fire against a person if you or they have reasonable grounds for believing he/she is currently committing, or is about to commit, an action which is likely to endanger your or their lives, or the life of any person, and if there is no other way to prevent this.

Firing without Warning

6. You and your men may fire without a warning if the giving of a warning or any delay in firing could lead to death or injury to you or them or any other person, or if the giving of a warning is clearly impracticable.

Warning Before Firing

7. If the circumstances in paragraph 6 do not apply, a warning is necessary before firing. The warning is to be as clear as possible and is to include a direction to surrender and a clear warning that fire will be opened if the direction is not obeyed.

Area of Operations

8. Under no circumstances are you or your men to enter Spanish territory or Spanish territorial waters for the purposes connected with Operation Flavius, nor are you or your men to fire at any person on Spanish territory or Spanish territorial waters.

Appendix C

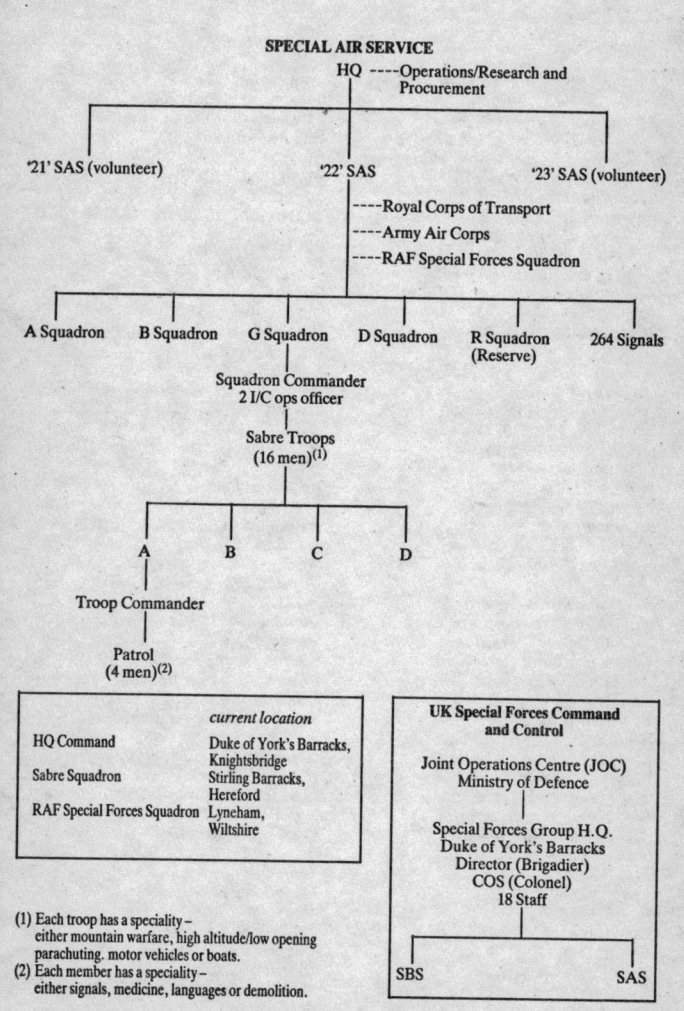

SPECIAL AIR SERVICE

HQ ----Operations/Research and Procurement

'21' SAS (volunteer)　　　'22' SAS　　　'23' SAS (volunteer)

----Royal Corps of Transport
----Army Air Corps
----RAF Special Forces Squadron

A Squadron　　B Squadron　　G Squadron　　D Squadron　　R Squadron (Reserve)　　264 Signals

Squadron Commander
2 I/C ops officer

Sabre Troops
(16 men)[1]

A　　B　　C　　D

Troop Commander

Patrol
(4 men)[2]

	current location
HQ Command	Duke of York's Barracks, Knightsbridge
Sabre Squadron	Stirling Barracks, Hereford
RAF Special Forces Squadron	Lyneham, Wiltshire

UK Special Forces Command and Control

Joint Operations Centre (JOC)
Ministry of Defence

Special Forces Group H.Q.
Duke of York's Barracks
Director (Brigadier)
COS (Colonel)
18 Staff

SBS　　　　SAS

(1) Each troop has a speciality –
    either mountain warfare, high altitude/low opening
    parachuting, motor vehicles or boats.
(2) Each member has a speciality –
    either signals, medicine, languages or demolition.

Index

Index compiled by Peva Keane